Sensor Testing

& Waveform Analysis

(Section 1)

By
Mandy Concepcion

All charts, photos, and signal waveform captures were taken from the author's file library. This book was written without the sponsoring of any one particular company or organization. No endorsements are made or implied. Any reference to a company or organization is made purely for sake of information.

www.autodiagnosticsandpublishing.com

By Mandy Concepcion.

Diagnostics Strategies of Modern Automotive Systems

By
Mandy Concepcion

All charts, photos, and signal waveform captures were taken from the author's file library. This book was written without the sponsoring of any one particular company or organization. No endorsements are made or implied. Any reference to a company or organization is made purely for sake of information.

www.autodiagnosticsandpublishing.com

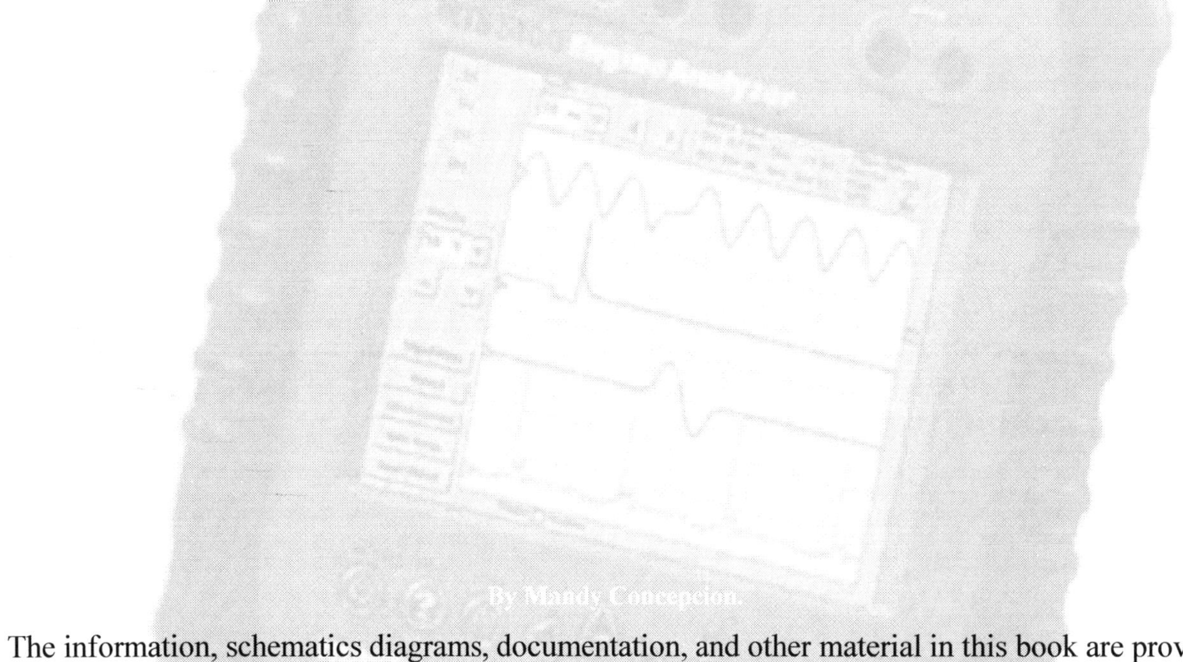

Made in the U.S.A.

AUTOMOTIVE SENSORS

Edition 4.0 Section 1– Automotive Sensors

In this section, we will deal with the vast array of automotive sensors found in today's vehicles. As most of this book, special attention is given to a practical diagnostic approach. At the same time, the procedures are also accompanied by the necessary theory to troubleshoot these sensors.

Section 1 Contents

O2 SENSOR TROUBLESHOOTING STRATEGY

THEORY OF OPERATION

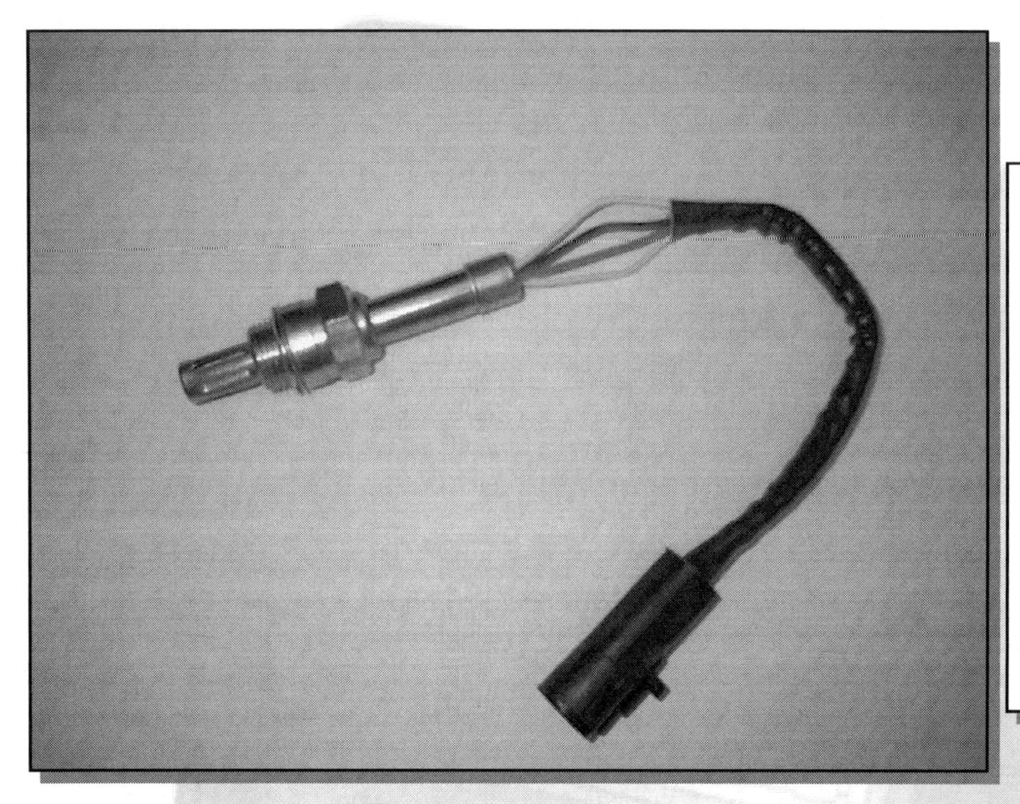

Fig 1 - 4 Wire OBD II O2 sensor. The two lighter wires are for the heater element, the gray wire is a signal ground, and the black is the O2 sensor signal output. DO NOT mistake the black wire for a ground.

The O2 sensor measures the oxygen content of the exhaust. The O2 sensor's sensing ability comes about by producing a small voltage proportionate to the exhaust oxygen content. In other words, if the oxygen content is low it produces a high voltage (0.90 Volts - Rich mixture) and if the oxygen content is high it produces a low voltage (0.10 Volts - Lean mixture). Although theoretically the O2 sensor should cycle between 0.00 volts and 1.00 volts, in reality it cycles between 0.10 volts and 0.90 volts.

A few key issues are very important in the analysis of O2 sensor signals.

- An O2 sensor will cycle between 0.10 to 0.90 or almost 1 volt.
- An O2 sensor has to reach the 0.8x Volts amplitude mark while at full operation.
- An O2 sensor also has to reach the 0.1x Volts amplitude mark while at full operation.

Full operation means the engine is fully warmed up, O2 sensor above the 600 deg. F. operating temperature, and no fuel or mechanical problems present.

- The O2 sensor must cycle at least **once per second,** which would show **3 cross counts** on the scan tool **PID**.
- **Silicone** is the leading cause of O2 contamination.
- It is easier for an O2 sensor to go from rich to lean than vise-versa.
- O2 sensors tend to fail on rich bias. In other words, they tend to shift their cycling to the upper side or rich side of the voltage scale.

- Contrary to what many people think, an O2 sensor **WILL NOT** cycle by itself. The O2 sensor cycle is a direct result of the ECM response to the changes in the mixture.
- Any time the O2 cycles and crosses the 0.450 volts mark, the system is in **CLOSE-LOOP**.
- Even though an O2 sensor is cycling and crossing 0.450 volts (ECM in close loop) it **DOES NOT** mean that it is working properly.
- O2 sensor operation is extremely important not only to keep HC & CO emissions low but also to the NOx as well.
- Proper O2 sensor cycling will determine the catalytic converter's efficiency. The catalytic converter needs the O2 sensor cycling at its proper amplitude and frequency for it to function at its maximum efficiency.
- An O2 sensor with a high voltage reading does not necessarily mean that the mixture is rich or high in fuel content. An EGR valve problem will send the O2 signal high as well.
- A GM O2 sensor signal stuck at 450 mV is an indication of an **open** O2 sensor circuit (signal wire) or **faulty O2 signal ground**. The 450 mV value (GM) is called a bias voltage and it is not the same for all manufacturers. Some manufacturers employ a dedicated O2 sensor ground. Such a ground lead is attached to the engine block or chassis and feeds an ECM O2 ground pin only. The O2 circuit is then grounded through the inside of the ECM electronic board by this ground wire. A loss of this ground would also put the O2 sensor signal at around 450 mV, which also makes it look like an open circuit. The same holds true for Chrysler, but these use a different O2 bias voltage, which is usually 2.00 to 4.00 volts.

A big misconception among technicians trying to understand O2 sensors is that they cycle by themselves. The O2 sensor just reads oxygen content in the exhaust, **THAT'S IT**. Excess oxygen in the form of regular ambient air will send the O2 sensor voltage signal low (under 0.450 volts) and lack of it will send the voltage signal high (over 0.450 volts). A stuck open EGR valve will create a lack of oxygen in the exhaust, since the re-circulating exhaust has all its oxygen already burnt . The ECM sometimes uses the O2 sensor to check for proper EGR operation and sets a code if necessary. So, be aware of the fact that a vehicle might be running lean because the ECM sees a rich O2 signal due to a defective (stuck open) EGR valve. Since the ECM sees a rich signal, it will try to correct with a lean command and try to lower the O2 sensor's high voltage signal.

CONDITION THAT AFFECT OPERATION

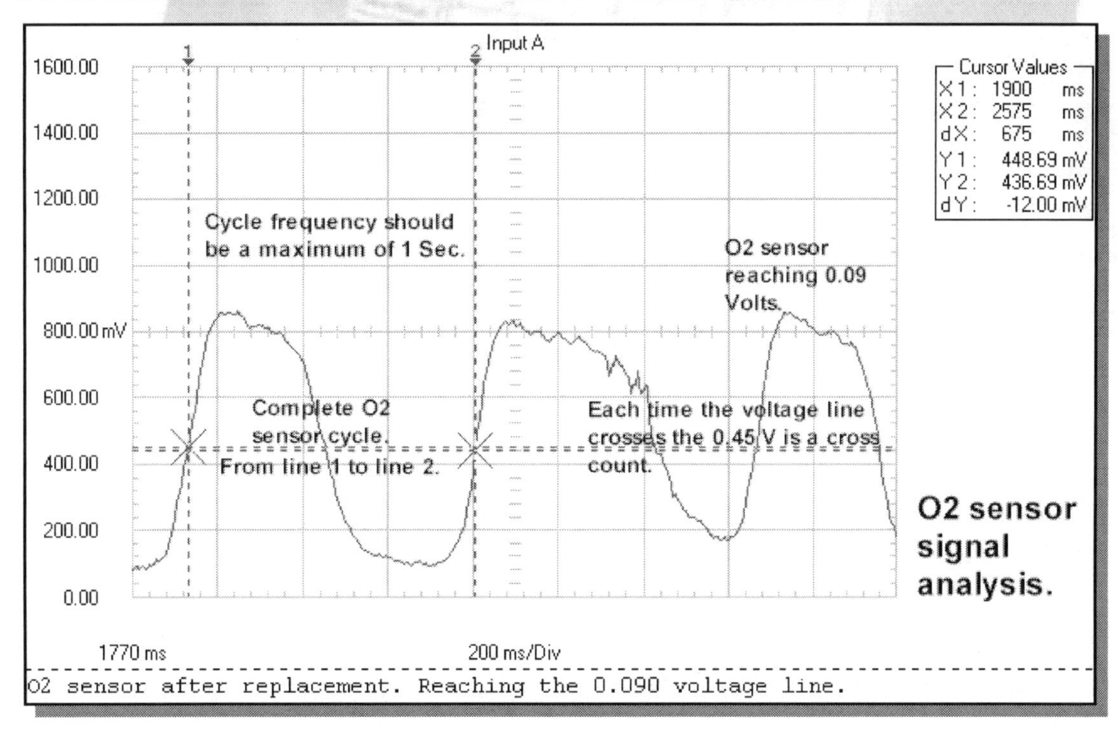

Fig 2 - Oxygen sensor signal analysis taking into account frequency and amplitude.

NOTE

WHEN PERFORMING O2 SENSOR CHECKS, IT IS IMPORTANT TO TAKE MEASUREMENTS AT IDLE AND 2000 RPM. BE AWARE THAT O2 SENSOR PRE-CONDITIONING IS IMPORTANT, EVEN ON THE NEWER STYLED HEATED O2 SENSORS. PRE-CONDITION THE O2 SENSOR BY RAISING THE ENGINE SPEED TO 2000 RPM FOR ABOUT 15 SECONDS OR SO. THE O2 SENSOR HAS TO BE ABOVE 600 ° F. TO BE ABLE TO OPERATE PROPERLY. LONG PERIODS OF IDLE TIME CAN RENDER A NON-HEATED OR OLDER O2 SENSOR TOO COLD FOR IT TO FUNCTION AT ALL. AT THE SAME TIME, DO NOT TRY TO FORCE A HEATED O2 SENSOR INTO OPERATION. AN O2 SENSOR WITH A FAULTY HEATER WILL GO INTO CLOSED-LOOP AFTER A GOOD WARM-UP SESSION.

After an engine has ran through its warm up period (O2 sensor has no effect on engine operation while the engine is cold), the ECM then looks for the O2 value. The 0.450 volts mark is considered almost universally as the midway point or crossover point for O2 sensor operation. If the signal is on the rich side (above 0.45 volts), then the ECM will answer with a lean command (reducing injector pulsation), or if the signal is on the lean side (below 0.45 volts) then the ECM will answer with a rich command (increasing injector pulsation). The amount of injector pulse correction is **proportional** to the voltage seen by the ECM at the O2 sensor signal wire. The higher the voltage the more the ECM reduces on-time to the injector. The lower the voltage the more the ECM increases the injector on-time. The ECM is constantly doing exactly just that, slightly increasing and decreasing injector pulsation. The constant adjustment is what gives the O2 sensor signal the switching appearance (sine wave) on the scope screen.

NOTE

The ECM's fuel pulse corrections performed constantly to the injector signal is called SHORT TERM FUEL TRIM (GM called it INTEGRATOR) and LONG TERM FUEL TRIM (GM called it BLOCK LEARN) on the scanner. FUEL TRIMS is the system's deviation of the BASE-INJECTION pulse. Analyzing LTFT and STFT is a great way to know a particular vehicle's fuel consumption trend or how well that vehicle has been performing with regards to fuel control. STFT and LTFT is the first thing to look for when assessing fuel control problems.

The fact that the O2 sensor signal is switching rich-lean-rich-lean also reveals that the ECM is controlling the injector pulsation and therefore that the system is in **close loop mode**. An ECM in full control (O2 sensor cycling) is said to be in close loop because of the **close-circuit action** of O2 sensor-to ECM-to injector pulse control then to O2 sensor and back to the ECM. The ECM must be in control at all times except during **warm up, WOT, power enrichment, and deceleration mode**.

The O2 sensor not only has to cycle, it also has to cycle fast enough (proper frequency) and wide enough (proper amplitude). At least **one cycle per second (1 Hz)** must be seen at the signal wire in order for the O2 to be considered good (not lazy). A one cycle per second will make the scope trace go across the 0.450 volts mark approximately 3 times, which the ECM recognizes as 3 cross counts. A slow O2 sensor will have a damaging effect on the catalytic converter and release excessive amounts of emissions to the atmosphere.

A cycle are the complete rich and lean crests of the O2 sensor signal, while crossing the 0.45 voltage point. Proper amplitude refers to the O2 sensor's ability to reach full rich (0.90 volts) and full lean (0.10 volts) when cycling. The higher the voltage seen at the O2 signal line the more the ECM reduces pulsation to the injectors. The lower the voltage seen at the O2 signal line the more the ECM increases injector pulsation. This is the reason why an O2 sensor that is not reading the mixture properly, at full amplitude and frequency, will actually misguide the ECM into a wrong fuel control pattern. Once the O2 sensor has reached its correct temperature of 600 ° F, look for an O2 signal cycle with the correct amplitude and frequency and it will surely indicate a perfectly operating O2 sensor.

COMPONENT TESTING

On early OBD II systems, the post catalytic converter O2 sensor has no effect on fuel control. The post catalytic O2 sensor was originally responsible for only monitoring catalytic converter efficiency. On most systems, the post converter O2 sensor signal should never mimic or follow the pre-cat O2 signal. That would indicate a defective or low oxygen storage capability at the converter . On early OBD II systems, the post-cat O2 sensor should show little or no voltage fluctuations on a scope waveform, since all the mixture fluctuations are being absorbed by the catalytic converter.

Stating around model year 1999, a new type of converter came on the market, called "Low Oxygen Storage Converter" or LOC. With an LOC, the pre and post O2 sensors cycle at the same rate. These converters are tested by measuring the lag-time between the two signals. A further development of this system is that the post converter signal is also used for A/F correction, but to a less extent.

These simple steps should be followed whenever testing O2 sensors.

1. Scan the vehicle for any O2 sensor codes and analyze the data stream PID. O2 sensor voltage should cycle normally with proper amplitude and frequency. An O2 sensor stuck at a fixed bias voltage is an indication of an open O2 circuit or lack of O2 sensor (dedicated) ground. If possible use a graphing multi-meter to analyze the O2 sensor data to determine any possible problems.
2. While reading the scan values, goose the throttle and observe for O2 sensor minimum and maximum values (0.1x volts to 0.9x volts). Although this is not a conclusive evidence of correct O2 sensor operation, it serves as a preliminary indication of proper operation.
3. Some automotive manufacturers employ a dedicated O2 sensor ground wire that is grounded somewhere at the engine block or chassis. A loss or rupture of this ground wire will render the O2 sensor useless. This ground wire feeds only the ECM's O2 sensor circuit. The main engine ground does not feed this type of O2 sensor circuit.
4. Verify the O2 sensor wire integrity. Most O2 sensors are biased and an open signal wire will give a reading of whatever the bias voltage is. Later model Jeep/Chrysler O2 circuits tend to be biased at around 2 or 4 volts, therefore, a constant reading of around 2 or 4 volts on a Chrysler is also an indication of an open circuit. In many of these cases, the ECM will put an **"O2 sensor High Voltage"** code.
5. Finally, verify for correct O2 sensor operation with a scope or graphing multi-meter. Check for proper amplitude and frequency. **Remember that the scanner O2 sensor readings are only interpreted values and may not show the real voltage reading. This is the reason for doing this final manual test.**

ACCELERATOR PEDAL POSITION SENSOR (APP) TROUBLESHOOTING STRATEGY

THEORY OF OPERATION

Modern automotive safety systems rely heavily on electronics to make it all work. The newer accelerator system, without any cable linkage, is no exception. This system is generically called **drive-by-wire** and it acquires different names depending on the manufacturer. Drive-by-wire integrates an electronically controlled throttle plate with a computer. Features such as cruise control, automatic idle adjustment, and other safety features such as automatic deceleration if the vehicle goes out of control are simply added as software add-ons. This makes it very inexpensive for manufacturers to keep up with newer technologies.

The accelerator pedal position sensor (APP) is of central importance to the drive-by-wire system. This sensor is designed to provide the ECM with accelerator pedal position as well as its rate of change or how fast the driver is pushing on the gas pedal. Besides providing the ECM with accelerator position, the APP is also associated with other modes of operation. One such mode is the reverse mode in which the ECM changes the acceleration pattern of the vehicle when backing-up and the selective cylinder-kill function if part of the APP sensor fails. In this last case, the ECM cuts injector pulsation every other engine revolution in the event of an APP failure. This reduces performance, but allows the driver to reach a repair shop.

The APP sensor is basically a potentiometer or variable resistor with a second potentiometer acting as a redundant sensor in case of a failure (some systems also use a third potentiometer). In other words, the APP is two or three position sensors in one. The first sensor (sensor 1) is the main input to the ECM for throttle plate control and the second sensor (sensor 2 or 3) is for redundancy. The ECM is constantly comparing all potentiometer readings against each other. In the event of a discrepancy the ECM sets a code and goes into a reduced performance mode (limp home mode). The ECM also provides a separate reference voltage and ground to each potentiometer, regardless of whether 2 or 3 pots of are used. If the reference voltage or ground is lost to one of the potentiometers the others are still able to function. This allows the driver to drive the vehicle, with limited performance, to the nearest shop. These series of potentiometers within the APP sensor are also wired differently for the sake of redundancy.

Understanding the system's wiring pattern is very important for correct diagnosis. A quick look at the wiring diagram will reveal the particulars to the system being worked on. On certain systems, it is common to see one of the APP potentiometer signal increase (low to high) when pressing on the accelerator, while at the same time the other potentiometer signal decreases (high to low).

The APP signal characteristics is dependent on how the APP sensor is wired. If the potentiometer's wiper rests (throttle closed) at ground, then the signal output increases as the throttle is pressed. And if the wiper rests at the 5 volt reference, then the signal decreases as the throttle is pressed. The important thing to remember is that regardless of how many potentiometers are within the APP, they all have separate reference voltage and grounds lines.

A throttle control actuator is also employed in drive-by-wire systems to do the actual throttle opening. The ECM responds to the APP signal changes by opening the throttle plate. Some systems also use a separate throttle control computer or throttle control module.

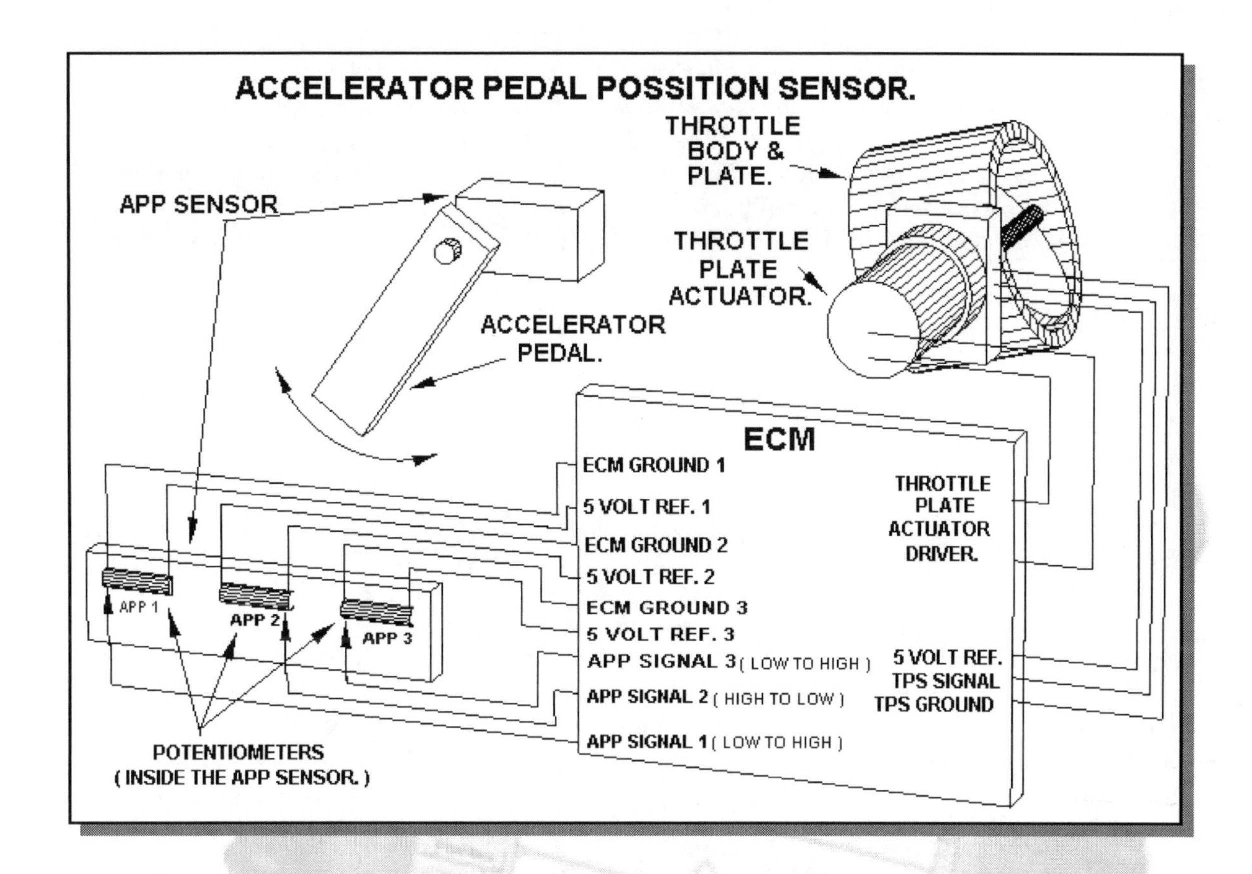

Fig 1 – APP sensor and throttle actuator circuitry. Notice the three potentiometers inside the APP sensor. Potentiometers wipers 1 & 3 rest at ground. This signal therefore starts at a low voltage and increases as the accelerator pedal is pressed. On the other hand the # 2 potentiometer wiper rests at the 5 volt reference. This signal starts high and decreases as the accelerator is pressed. The three signal readings change simultaneously while pressing on the accelerator pedal.

The throttle control module (if separate) will always work in conjunction with the engine control module in order to control the vehicle's acceleration. Both modules communicate with each other via a serial data line. In older systems there were multiple sensor connections from and to each of these modules. But newer and faster ECMs have made it possible to rely on data line communications and not hard-wired systems. As the APP signal changes, the ECM commands the throttle actuator motor to open the throttle plates a certain amount. The throttle actuator assembly TPS sensors then relay the throttle opening back to the module. The APP signal is constantly being compared to the throttle actuator assembly position sensors. Any discrepancy between the APP and the TPS sensors will set a code, with the system also going into limp-in-mode. The ECM flags a discrepancy when the deviation or difference between the APP and the TPS signal goes above a maximum pre-programmed amount. All drive-by-wire systems use dual TPS sensors. Again this is all done for redundancy.

NOTE

It is important to know that under no circumstances should the throttle plates be forced open with a screwdriver. Some techs have tried this using screwdrivers or pliers to accelerate the engine, as with older systems (non drive-by-wire). This action will cause a discrepancy between the APP signal and the throttle actuator TPS sensors causing the vehicle to go into limp-in-mode (loss of performance). In most cases, the use of a dedicated OEM scan tool is needed to reset this condition and re-learn a new throttle actuator position adaptive memory.

In drive-by-wire systems, idle speed is achieved in two ways. Some systems use a regular IAC valve to control idle speed. There is no basic difference between the old and this new IAC valve. The only exception is in the internal ECM programming, which takes into account the differences in the throttle actuator.

The second way to control idle speed is by the ECM using the actual throttle actuator to slightly open the throttle plates. This action, in conjunction with the engine speed input, and the TPS signal determines the idle speed. As a side note, it should be pointed out that in older (non drive-by-wire) systems the throttle plate and bore carbon deposits were the cause of a good deal of idle air/fuel mixture problems. However, in drive-by-wire systems, dirt and carbon deposits could render the engine inoperative by making it unable to idle properly. It all goes back to the fact that any discrepancy between the APP and the TPS will cause a variety of different problems. In this example of an engine with a dirty throttle body, the carbon and dirt creates an air restriction causing the ECM to increase or change the throttle plate opening. The ECM has pre-programmed in its internal memory the maximum throttle opening at idle possible. If this limit is reached, (because of the carbonized throttle body) a trouble code is set in memory and the system goes into limp-in-mode. A throttle body cleaning will correct any of these problems.

CONDITIONS THAT AFFECT OPERATION

As stated before, dirt and carbon deposits in the throttle plates have a negative effect on throttle control. The ECM will always try to adjust or compensate for the air restriction in the throttle body. This excessive throttle opening creates the APP to TPS signal discrepancy, triggering the ECM into setting a code.

The APP dual or triple potentiometer signal must also be within proper specifications. An improperly adjusted (if adjustment is available) APP sensor at the rest position will be picked up by the ECM, as is a bad or erroneous signal. Also, the two or three APP signals must be without discrepancies within each other. If the main signal fails the ECM then looks at the other signals for proper accelerator pedal position. This will set a code and performance will be reduced.

The throttle actuator is an actual electric motor (stepper or DC motor) that is energized by the ECM or throttle module in response to the APP signal. The actuator has a set of gears and springs which enables it to work the throttle plates. Any binding of the gears or breakage of the springs will create an immediate discrepancy between the APP and the throttle TPS signals. The same applies to a defective throttle actuator motor itself. Again, a discrepancy beyond the pre-programmed acceptable limit will set a code and the ECM will go into reduced engine performance. On certain occasions, if the TPS dual signal is completely lost the engine will shut down. Hence the use of two TPS signals with independent voltage reference and ground. Again, all done for redundancy.

COMPONENT TESTING

An invaluable tool in diagnosing APP and drive-by-wire system problems in general is the scanner. An after market or preferably an OEM scan tool analysis, with all the parameters (PIDs) necessary for proper signal monitoring makes for a quick preliminary testing of the system. A quick analysis of the APP PIDs will indicate if the potentiometer signals are out of specifications and any possible signal discrepancy can be picked up by analyzing the different PIDs. A careful observation of the APP signals, while slowly pressing on the accelerator cable, will reveal a faulty APP sensor (provided that the scan tool is fast enough, as with some OEM scanners) . Otherwise, a multi-channel VOM or scope should be used. It is also worth knowing that most systems will put out **throttle position ERROR PID**. Watch carefully for this parameter, since an error flag will reveal an APP to TPS discrepancy.

Despite today's faster and better scan tools, the second step to this procedure should always be followed by a manual electrical check.

- The first step is to prove the APP voltage reference and ground circuits. These circuit are provided by the ECM independently of each other, for redundancy.
- Proceed to the throttle computer (look up the proper name according to the manufacturer) and uncover the wiring at the connector itself.
- Check the signal voltage output by using a multi-channel VOM or oscilloscope. Connect each channel to the two or three potentiometer output signals. The need for a multi-channel scope is becoming more apparent as more sensors will be added to future vehicles.
- With the scope connected, check the output signals with the APP sensor at rest. Compare to proper specifications. Then slowly press on the accelerator pedal and observe for any glitches or sudden drops in signal voltage. This procedure is somewhat similar to checking a TPS sensor. The difference is that it is a dual or a triple TPS (depending on manufacturer).
- Notice the correlation of all the signals to each other. If one of the potentiometer signals output is off calibration, the ECM will set a faulty code.
- If the APP sensor passes the test, disconnect the scope/VOM from the two redundant potentiometers. Then, connect the second and third channels to the dual TPS at the throttle plates. Leave the first channel connected to the primary potentiometer. With the engine running accelerate the vehicle slowly. Observe the TPS outputs for a smooth increasing signal, as during a normal TPS check. This step will indicate any throttle actuator binding or TPS problems. The test will also uncover a possible ECM transistor driver problem or an electrically faulty throttle actuator motor.

NOTE

If an OEM scan tool is available and the manufacturer has bi-directional control of the throttle actuator, perform an increasing throttle opening command and observe the TPS output on both the scope and the scanner PIDs. Always be on the alert for any APP to TPS discrepancy that might send the system into limp-in-mode.

- A couple of other scan tool data PIDs are also helpful in diagnosing this system. If the system uses the throttle actuator for idle control, look for an out of balance air/fuel ratio mixture. The long term and short term fuel trims (LTFT & STFT) are helpful PIDs when it comes to diagnosing air/fuel ratio problems. Provided that the vehicle has no vacuum leaks or fuel restrictions, an out of adjustment or faulty TPS will send the wrong signal reading to the ECM. A maladjusted TPS could make the ECM react as if the throttle plates were in a different position than they really are. The net effects will be the ECM increasing throttle plate opening (this will create a lean condition), or reducing throttle opening (this action will create a rich condition or a possible engine stall).

By studying the system PIDs carefully before jumping into a long diagnostic routine, it is possible to pin point the fault faster and easier. **Knowing the system is the first rule to remember in modern vehicle diagnostics.**

CAM & CRK SENSOR TROUBLESHOOTING

THEORY OF OPERATION

CRANKSHAFT POSITION SENSOR

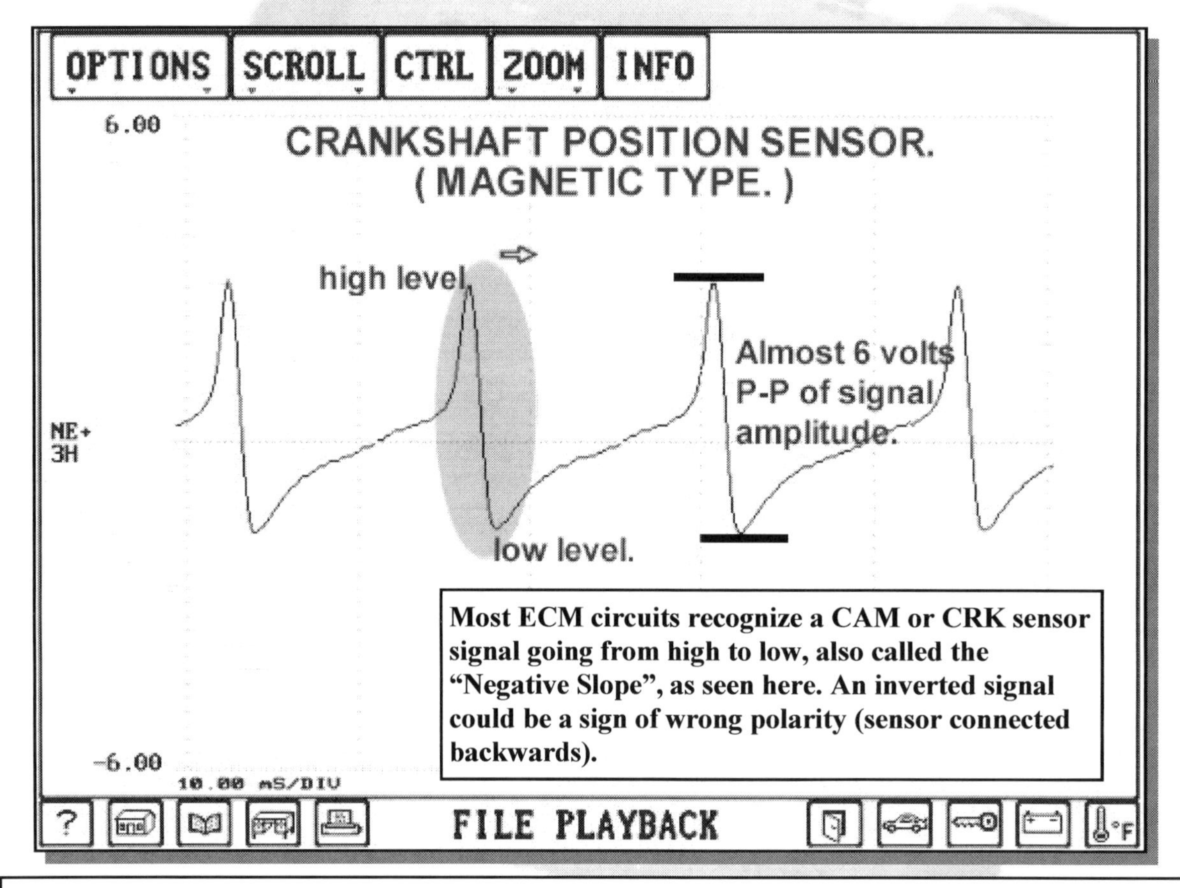

Fig 1 – CRK sensor analysis. Here, an actual magnetic CRK sensor signal shows an amplitude of about 6 volts peak-to-peak. The amplitude recognition threshold voltage is critical for the ECM to be able to recognize the signal. The signal amplitude should always be over the recognition threshold voltage.

The crank (**CRK**) sensor signal is probably the most important signal in a modern automotive engine control system. This signal provides the ECM with crankshaft speed and position, as well as a cylinder # 1 reference point. There are various names given to this signal. The distributor reference, CRK signal, CAS, PIP, etc (depending on the manufacturer). **The way this signal reaches the ECM** will affect the approach that is taken to a proper diagnostic procedure. By analyzing the **signal path** to the ECM using a wiring diagram and an oscilloscope, the correct diagnostics determination can be made.

CAMSHAFT POSITION SENSOR

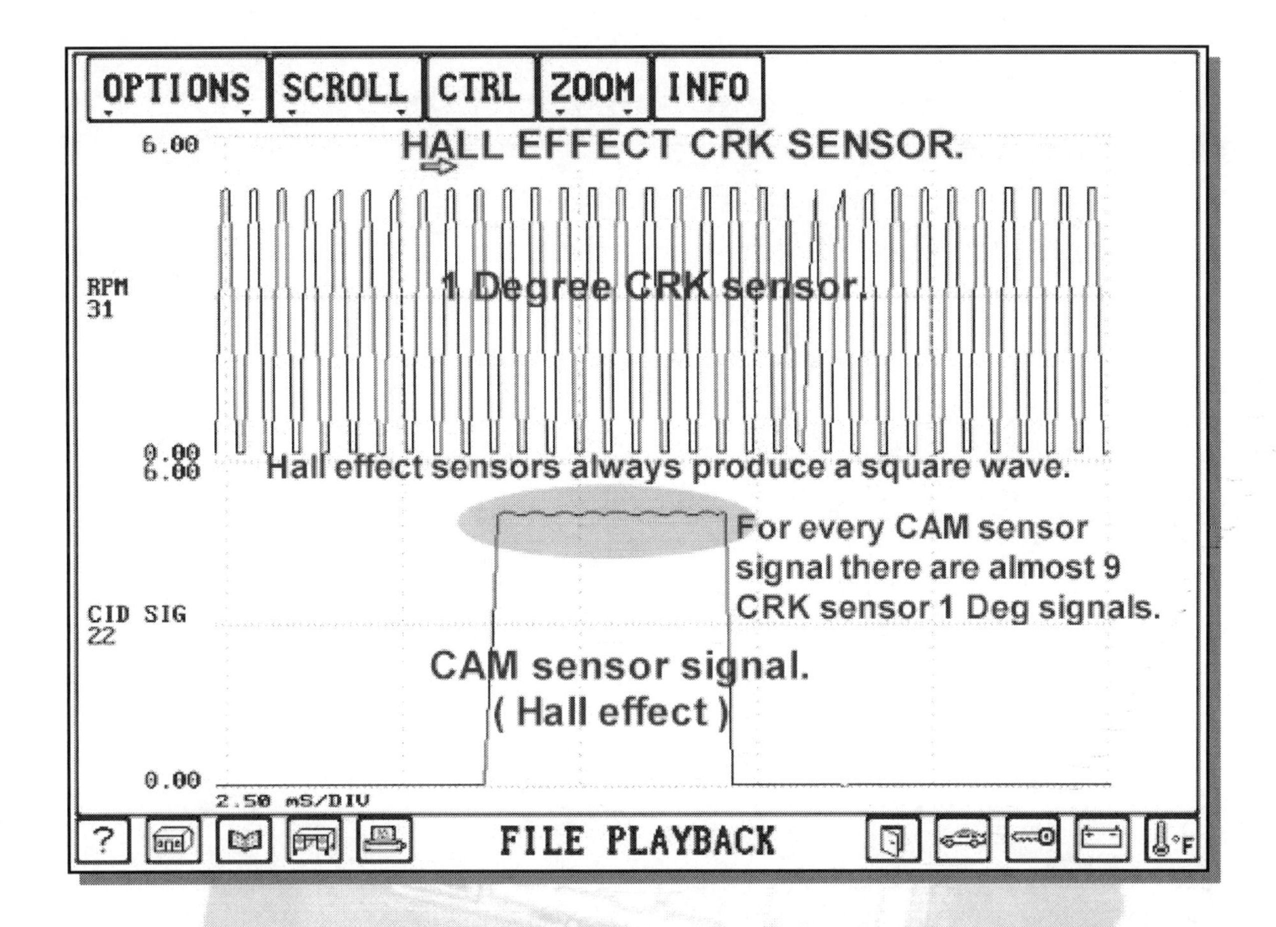

Fig 2 – CAM sensor hall effect signal analysis. Here the bottom CAM signal is compared to the Upper CRK signal. The CRK signal actually toggles about 9 times for every CAM output toggle. Hall effect and optical sensor signals are not affected by amplitude problems, since their output amplitude is always constant.

The **CAM** sensor signal is found on systems with sequential fuel injection, in which the ECM triggers the injectors independently instead of in group mode as in older systems. The CAM signal is also called CID, TDC, etc, depending on the manufacturer. The CAM sensor provides the ECM with camshaft position so that it can determine the correct injection and ignition sequence. Some systems (with distributors) do not need the CAM sensor to start the vehicle, and can simply start in non-sequential mode. However COP and most DIS systems do need the CAM sensor so that the ECM can determine the position of cylinder # 1 TDC on compression stroke and fire the correct coil.

The relationship between CAM and CRK signal is very important for proper ignition sequencing to occur. A stretched or jumped timing belt/chain will create severe engine performance problems on DIS/COP systems, since the ECM doesn't know when to trigger the coils. On other systems the ECM will shut down ignition entirely if it sees a discrepancy between these two signals.

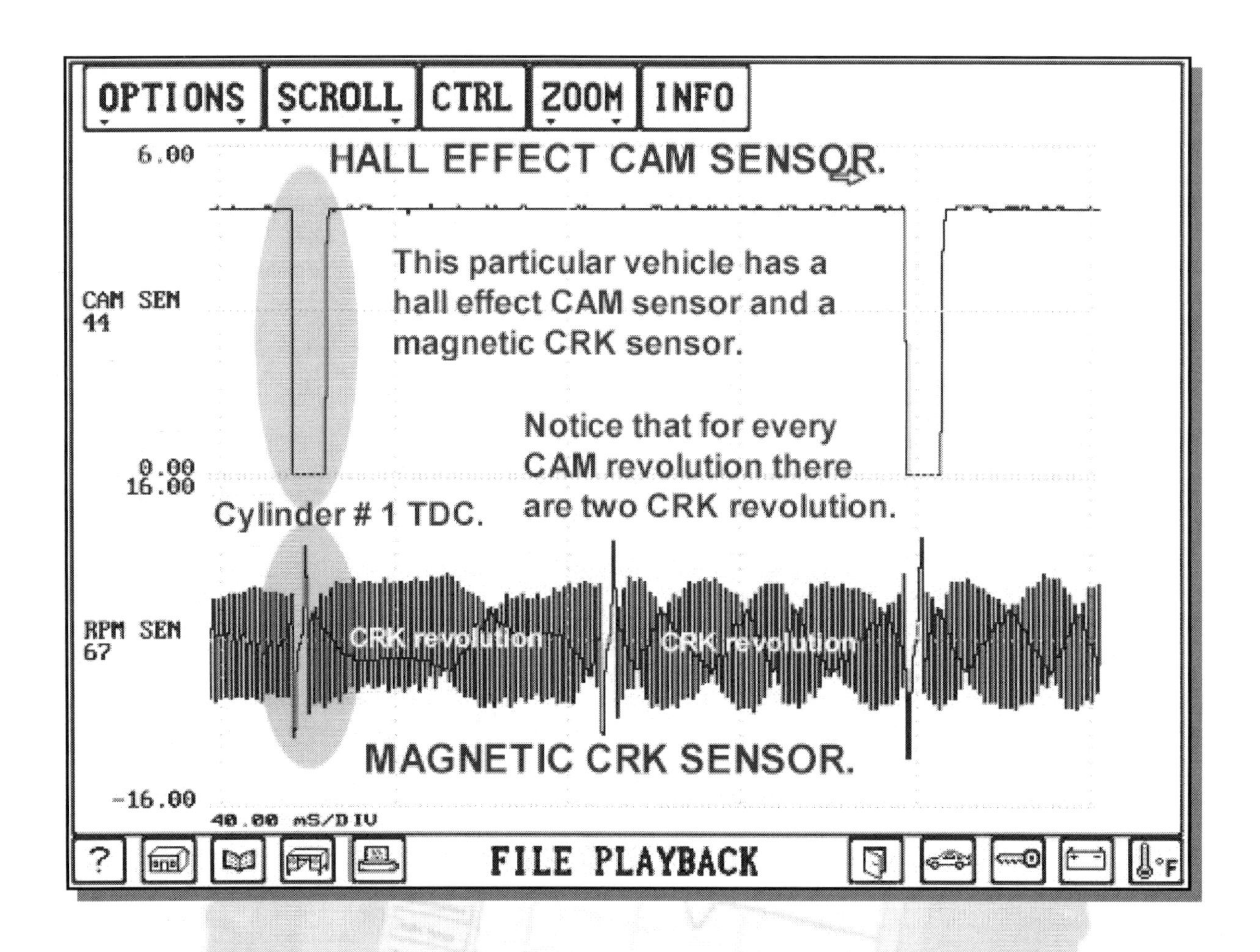

Fig 3 – This vehicle employs both a magnetic CRK sensor and a hall effect CAM sensor. The synchronizing of these two signal is critical for proper engine operation. Some systems will actually shut the ignition off if these two signals are out of synch.

CAM and CRK sensors come in four different varieties: **MAGNETIC, HALL EFFECT, OPTICAL AND MAGNETO-RESISTIVE.**

- The **magnetic** sensor actually produces its own signal. It is in essence a small generator. A coil winding inside the sensor picks up the magnetic fluctuations from the vibration damper or the flywheel (or both in some cases). A toothed reluctor wheel on either the damper or flywheel induces a voltage signal to the sensor. Magnetic sensors work on the principle of induction, which states that a metal object or magnet when placed across a coil winding will induce a current on that coil. Magnetic sensors are heavily dependant on the **air gap** between the sensor and reluctor wheel, and on the **speed of rotation**. The air gap has to be set as close as possible without touching the reluctor, and the engine cranking rotational speed has to be fast enough to produce the right signal amplitude. It is common to see vehicles that will not start due to a defective starter that is cranking the engine slower than normal . Systems that employ a magnetic sensor also have a threshold voltage, which is the voltage value at which the signal is first recognized by the ECM. Most distributor pick-up coils are of the magnetic type although hall effect distributors pick-ups are also found on some systems.

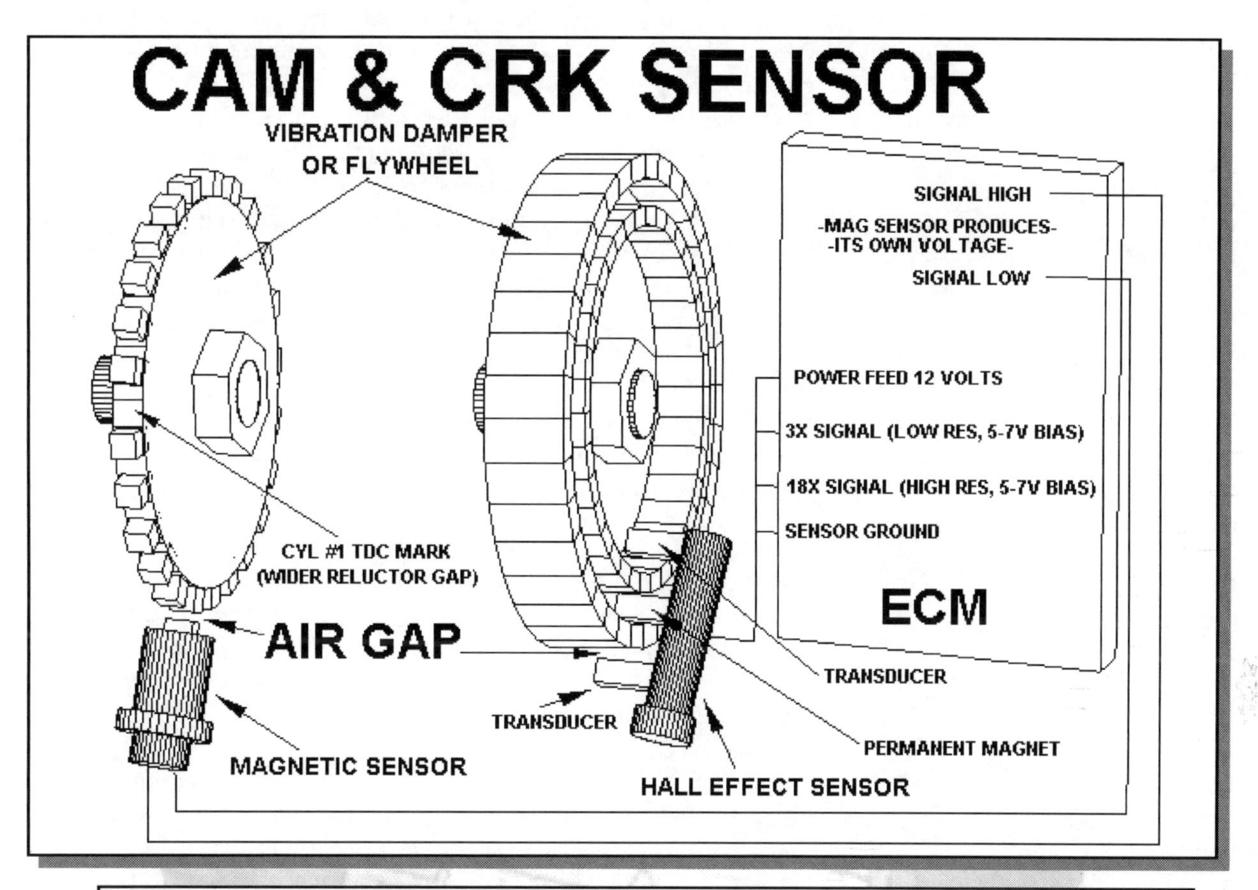

Fig 4 – Diagram shows the differences between the magnetic and hall effect sensor.

Once the signal reaches this pre-programmed voltage the ECM recognizes the signal and will act upon it (pulse the injector, etc). Magnetic sensors are usually shielded or with its wires twisted to prevent **electromagnetic interference**. On some systems the ECM provides a small **bias voltage** for diagnostics purposes. If the ECM sees a problem with this bias voltage, it will set a code for either a shorted or open circuit. Special attention should be paid to the polarity of these sensors. They are **polarity sensitive**. If for whatever reason the polarity (wires) is inverted, the vehicle will not perform properly or will not run at all.

- The **hall effect** sensor requires its **own voltage** and contains a switching transistor within the sensor casing. This type of sensor needs a voltage supply, reference voltage and a ground to operate. Transistors are electronic switches that turn ON or OFF when a current is applied to one of its three leads (**Base lead**). The sensing semi-conductor device or miniature coil in a hall-effect sensor is tied to the base lead of this internal transistor.

When the triggering mechanism (reluctor wheel) comes close to the hall effect sensor the magnetic lines cut across the sensing semiconductor device, which triggers the small internal transistor. This internal transistor then toggles the reference signal between ground and reference voltage. Hall effect sensor outputs a square wave signal simply because all they do is toggle their reference voltage to ground. In essence they are magnetic sensors, with an added internal switching transistor so that the sensed signal goes to the **base lead** of the internal transistor to trigger it instead of straight to the ECM, like a regular magnetic type sensor. Some hall effect sensors actually employ their own permanent magnet within its casing. This variation uses a shutter type triggering wheel that breaks across the magnetic field. The momentary interruption of this magnetic field is what triggers the base of its internal transistor. Regardless of what hall effect sensor variation used, they all output a square wave. Hall effect sensors are not affected by **slow engine cranking speeds.** They will simply toggle the reference voltage to ground, regardless of cracking speed.

- The **optical** sensor uses a principle somewhat similar to the hall effect sensor, but instead uses light as its triggering method. Optical sensors are light activated devices. These sensors use an LED (light emitting diode) as their light source, and a phototransistor as their triggering component. Optical sensors always have a shutter disk with small holes. Due to the more sensitive nature of the phototransistor, these holes are fairly small and can detect tiny amounts of engine speed fluctuations. Optical sensors are much more exacting in their operation and are able to detect very small engine variation problems much faster than any of the other two of sensor variants. Optical sensors also put out a **square wave**. They need a **supply voltage** and **ground** to feed the LED light source and phototransistors, as well as a **reference voltage**. The shutter wheel passes between the LED and the phototransistor; and as this shutter wheel turns, it momentarily breaks the light beam emitted by the LED. This light beam breaking action is detected by the photo-transistor, which instead of having a base lead has a small lens or eye that is always looking for the light source. The action of the shutter wheel breaking the light source also triggers the phototransistor, which in turns toggles the reference voltage to ground. Optical sensors may also have two LED light sources. One for the 360 ° of crank rotation and the other with 4-6-8 slots to denote each cylinder position depending on the amount of cylinders on the engine. It is fairly common to see dirt and oil contaminate the small holes on the optical triggering wheel and cause erratic or no signal output at all. Neither optical or hall effect sensors are affected by **slow engine cranking speeds.**

- The newer styled **magneto-resistive** sensor is yet another derivative of the hall-effect sensor. This sensor also puts out a square wave, but with one fundamental difference. Magneto-resistive sensors **DO NOT** ground their reference voltage. They are constructed with two internal sensing pick-up devices one besides the other. When the reluctor wheel tooth comes into proximity with the sensor, the first of the two sensing pick-up devices will trigger the base of the transistor and toggle the output signal high (i.e. 5 volts). A split second later, the second of the two sensing pick-ups will then toggle the output signal low (0 volts) or ground. This sensor uses the **leading and trailing edges** of the reluctor tooth to output a square wave. The leading tooth edge toggles the sensor high and the trailing edge toggles it low. The output is a regular square wave. Magneto-resistive sensors are also not affected by **slow engine cranking speeds**.

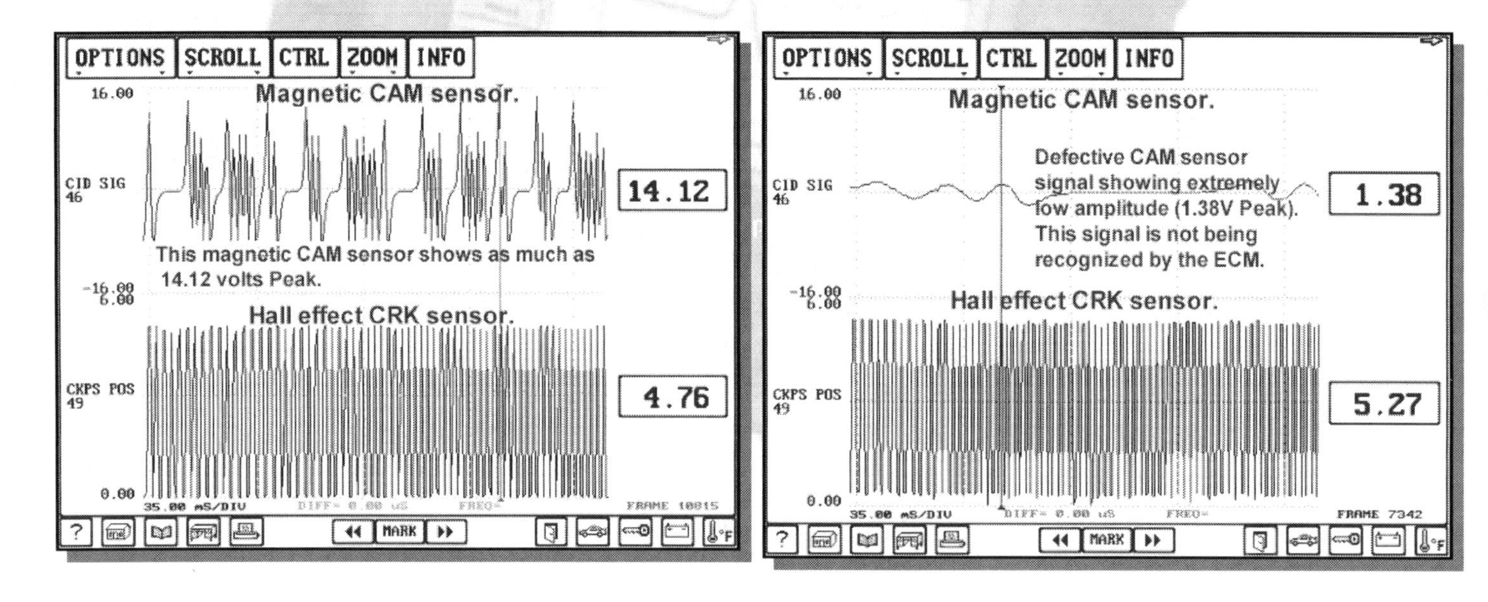

Fig 5 – This figure shows the difference between a good and bad magnetic CAM signal. The signal at the left shows proper signal amplitude of 14.12 volts Peak while the right signal shows a defective CAM sensor with a low amplitude of 1.38 volts. In this particular case the CAM sensor was at fault, however improper sensor air gap or low cranking speed will also cause this type of signal amplitude problem.

CONDITIONS THAT AFFECT OPERATION

The following conditions should be used as guidelines affecting all CAM & CRK sensors mentioned here. It is always important to determine the specific vehicle operation before making a diagnostics decision. Keep in mind that the way the CAM or CRK signal reaches the ECM will determine the diagnostic route to follow. **These signals will either go to the ignition module first then to the ECM or just straight to the ECM**. If a CAM or CRK code is set, careful consideration should be given to the particular vehicle strategy. A signal that first goes to the ICM and is not reaching the ECM could be due to it being shorted/open circuited at the ICM. Furthermore, on most of the sensor-ICM-ECM type of systems the actual hall effect voltage reference is provided by the ICM itself. These smart ICMs make all the decisions after processing the actual CAM/CRK signal and only then send a reference position signal to the ECM. A quick glance at the wiring diagram should be the first step. Learn and study the particular system before attempting to perform a diagnostic.

Magnetic sensor signal output strength (amplitude) is very dependant on the air gap between it and the triggering mechanism (reluctor wheel), and also the speed of engine rotation. The air gap usually comes out of adjustment over time due to engine vibration. Although the air gap on most magnetic sensors is not adjustable, dirt and metal filing tend to stick to the tip of the sensor and cause air gap sensing problems. A simple cleaning sometimes fixes the problem. Engine cranking speed is greatly affected by battery and starter condition. A slow cranking speed problem might make the vehicle not start at all. The lower cranking speed will also lower the sensor's signal amplitude. Internal sensor coil condition is also a main cause of magnetic sensor failure. Water and moisture get into the casing and corrodes the sensor's internal coil.

Hall effect sensors are fairly unaffected by engine cranking speed problems. They can still output a square wave even if the engine is turned by hand. Air gap and dirty sensors is also a main problem for hall effect sensor, as well as internal degradation due to corrosion and vibration.

Optical sensors main ailment is dirt in the optical shutter wheel. Since these sensors are much more sensitive, anything that interferes with the light beam will also affect the output signal. These sensors are not affected by low cranking speeds and they have no air gap to contend with. However, a warped optical shutter wheel may also render the sensor useless.

Magneto-resistive sensors are not affected by slow engine cranking speeds either. They are however extremely sensitive to signal noise created by out-of-adjustment air gaps and dirty sensor tip. Because of their signal noise sensitivity, clean sensor tips are a must with these sensors.

COMPONENT TESTING

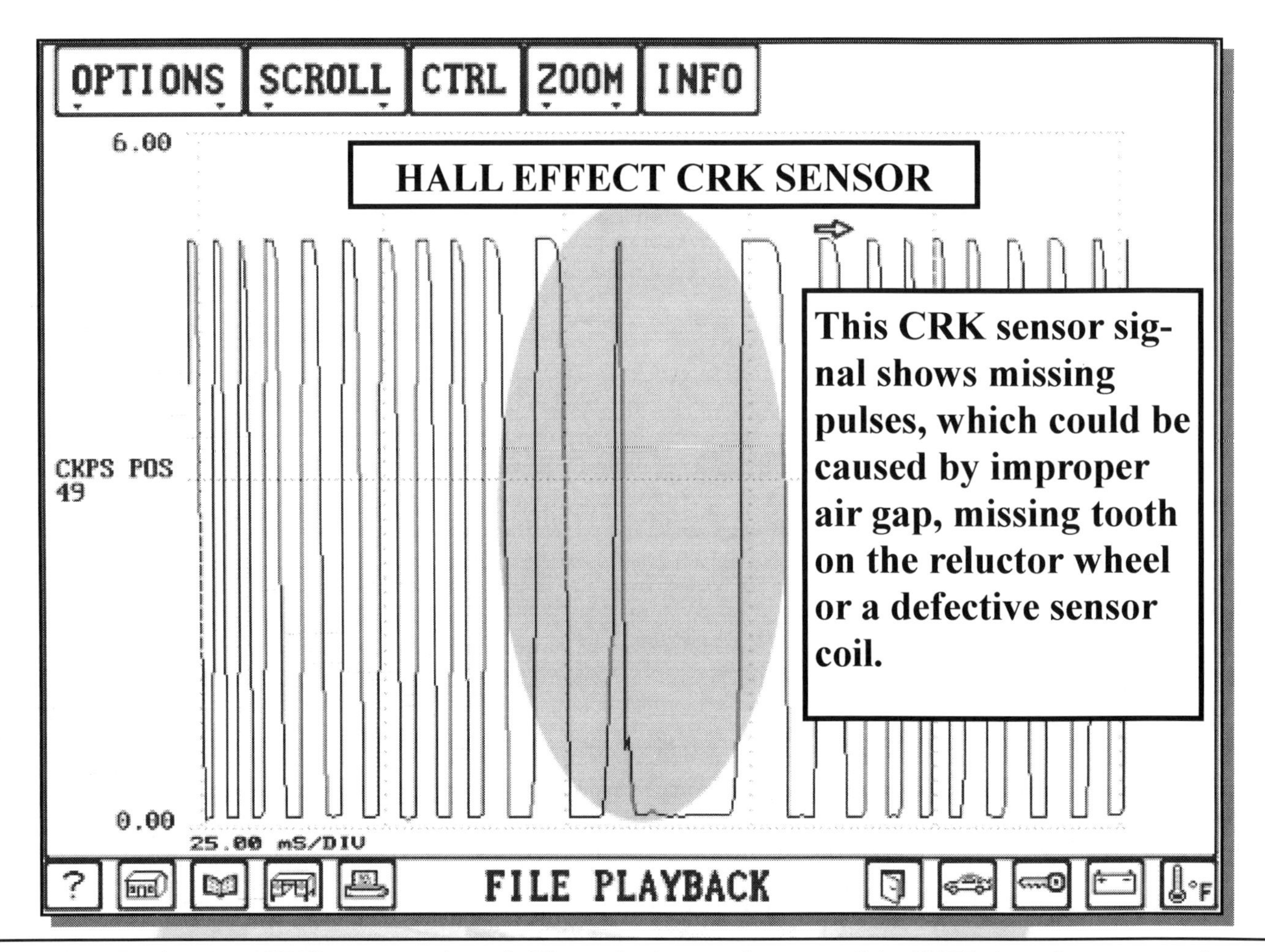

Fig 6 – Defective CRK sensor signal. The missing pulses on this CRK signal was causing an engine misfire. The CRK signal is the main input to the ECM for crankshaft position and speed. Most OBD II systems use a 1 deg signal to detect engine misfires, which would look similar to this signal above.

MAGNETIC SENSORS:

- The first step in testing a magnetic sensor is to scope the signal output for proper amplitude and frequency. The quality of the wave should be consistent as well.
- No sudden drops should be present. Remember that magnetic sensors produce their own voltage. A low amplitude problem is either the result of improper air gap or low starter cranking speed. It is even possible for an ECM to put out a CRK or CAM code due to a bad starter or battery. Always make sure that the starting and battery systems are up to specs.

HALL EFFECT & OPTICAL SENSORS:

- Because of the similarities between these two types of sensors the procedure to test them is much the same. Always check for proper **feed voltage**, **reference voltage** and **ground** first. These particular voltages vary between different manufacturers. The ECM provided ground is the second point to check with this type of sensor. Once it is verified that power and ground are present the sensor should toggle the reference voltage. Remember that low cranking speed does not affect this type of sensor.

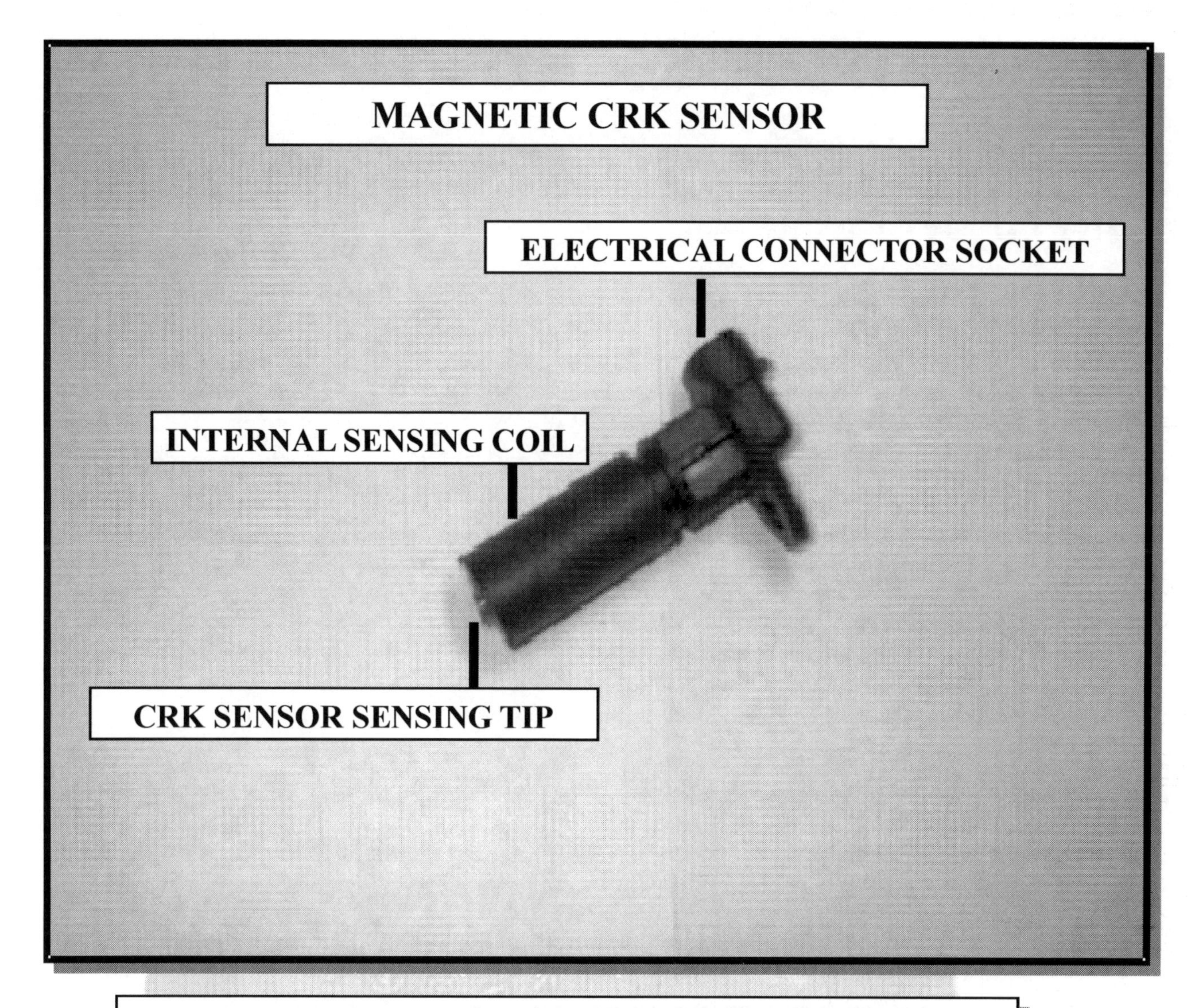

Fig 7 – Magnetic CRK sensor illustration showing its internal construction.

MAGNETO-RESISTIVE SENSORS:

- These are basically tested in the same manner as the hall-effect type sensor with an added difference, they do not toggle a reference voltage to ground. Always check for supply voltage and ground. If these two are present and the sensor is operational, there should be a square wave signal output.

As explained before, study the signal path from the sensor to the ECM. And, always consider whatever stands between this sensor and the ECM as possible suspects. A thorough understanding of the system being worked on is of crucial importance to proper diagnostics. This article simply tries to dissect and summarize the workings and procedures needed to perform a proper diagnostics.

ENGINE COOLANT TEMPERATURE SENSOR TROUBLESHOOTING (ECT)

THEORY OF OPERATION

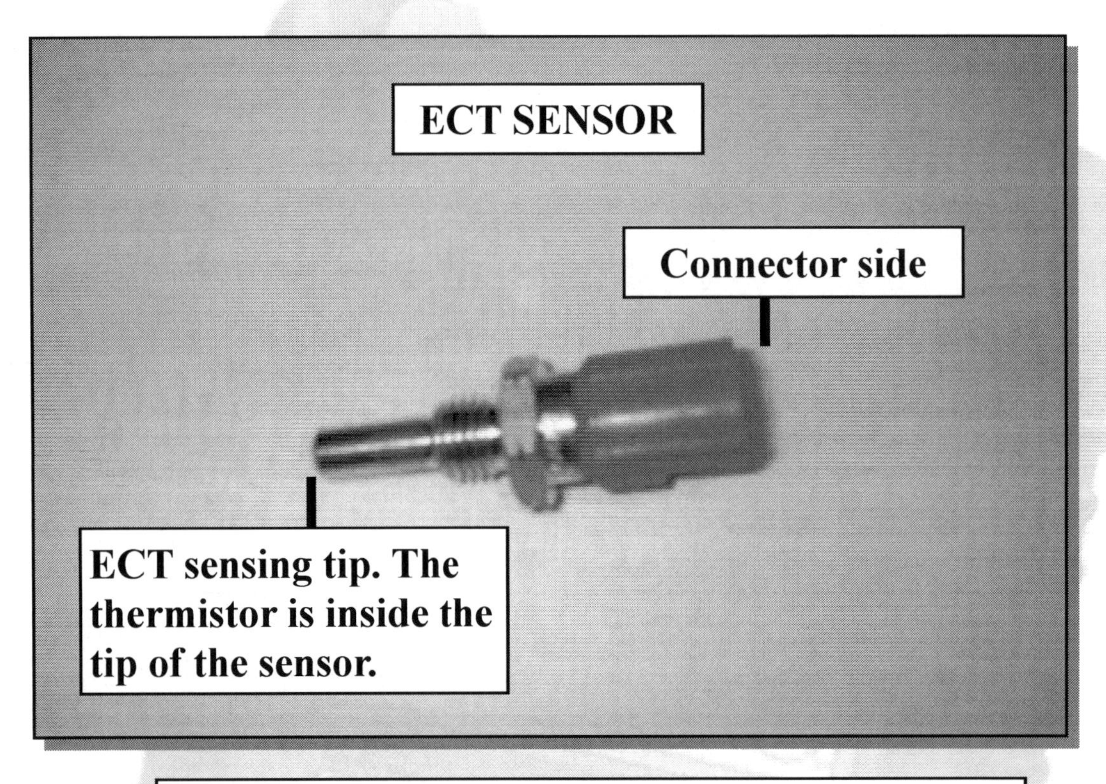

Fig 1 – Engine coolant temperature sensor (ECT).

The engine coolant temperature sensor (ECT) is a device that changes resistance as temperature changes. Its operating characteristic is linear, which means that when plotted on a chart the plotted line is straight. Inside the sensor is found what is called a thermistor, which is an electronic temperature sensitive variable-resistor. The ECT (thermistor) is a negative temperature coefficient sensor. This means that as temperature goes up resistance and voltage goes down or vise-versa. The ECT sensor receives a 5.00 volts reference voltage from the ECM. The ECT works by changing its internal resistance according to coolant temperature and therefore also changing the voltage drop across itself.

Some manufacturers use a dual scale ECT signal determined by **two different pull-up resistors inside the ECM**. A high (3.60 K Ohms) and low (345 Ohms) impedance circuit is provided internally by the ECM for the 5.00 volts reference. At low coolant temperatures the ECM uses one of the pull-up resistors (3.60 K ohms). As temperature increases and crosses over a certain temperature value (around 120 deg. F) the ECM switches the 5.00 volt reference internally to the other pull-up resistor (345 Ohms), so that in effect you'll see 2 volts at 86 deg. F and again 2 volts at 194 deg. F.

This is done to improve ECM control on a hotter engine. In essence, when plotted on a chart this voltage signal has a dual line curve. The actual ECT sensor however works the same way as any other coolant sensor. The difference is that in the dual resistor type system, the ECM internally switches voltage scales after the 120 ° F temperature mark is reached resulting in the dual charted voltage curve.

Important! The above voltages are fairly generalized. Variations will be found among different manufacturers, but the general operation is the same.

The ECT sensor comes as either a 2 or 3-wire type. The vast majority of ECT sensors are of the 2-wire type. These are connected on one side to an ECM provided sensor ground and the other side to a reference line (usually 5.00 volts), also from the ECM through the dropping resistor inside it. As temperature goes up the resistance of the ECT (thermistor) goes down and so does the voltage drop across it. The resistor in series with the ECT inside the ECM is provided for current limiting and as a voltage divider. This way in the event of an ECT wiring short the ECM would not get damaged. This is why on many occasions, it is possible to simply jump the two ECT terminals to determine cooling fan operation or injector control reaction. The 3-wire type ECT sensor works the same way as the 2-wire, it just has an extra thermistor inside to provide a signal to the temperature gauge or separate ignition module. In essence the 3 wire ECT sensor has the ECM coolant sensors and the gauge temperature-sending unit built inside the same sensor casing. A wide user of the dual ECT sensor is Mercedes Benz and some other Euro manufacturers, whereby the second ECT sends a signal to the ignition module as well. With this systems, the dual thermistor ECT sensor feeds both the ECM and the ICM or the ignition control module. Most early European systems had separate fuel and ignition modules.

There are some older systems of the early 1980's that used a dual thermistor ECT sensor, with only two signal wires coming out of it. This rare type of ECT sensor had the two thermistors tied together. The two connector prongs were connected directly to each side of the thermistors, while the center thermistor-to-thermistor connection was tied to ground, though the sensor's body casing. This ECT sensor was widely used by some early European systems, like Renault and Peugeot. The dual thermistor ECT sensor should not be confused with the ECM's internal dual pull-up resistor systems, having dual scales. The dual thermistor ECT sensor is simply two ECT sensors combined into one casing.

The ECM uses the ECT input to determine various component operation and engine control modes. It uses the ECT signal input to make calculations for the following.

- Fuel delivery or injector pulse-width.
- Cold start enrichment mode.
- Cooling fan operation.
- Determine open and close loop initiation.
- Idle control (IAC) operation.
- Ignition timing corrections.

On some systems, the ECT is disconnected to adjust timing, as in Chrysler systems.

The ECT is also used by the ECM to control the operation of the following components.

- Delay EGR, TCC and canister purge operation on a cold engine.
- Knock corrections depending on engine temperature.
- Some systems use the ECT input to control transmission shifts points and quality.

CONDITIONS THAT AFFECT OPERATION

Any condition that changes the resistance of the ECT circuit other than a temperature change is bound to have an adverse effect on vehicle performance. High CO levels, engine flooding, faulty cooling fan operation, etc are a result of a defective ECT sensor or circuit. An inoperative cooling system (stuck thermostat, inoperative cooling fans, etc) will also change the resistance of the thermistor inside the ECT sensor causing an engine performance problem.

[There are two possible scenarios as far as the ECT signal is concerned. A fault leading to excessive resistance or a lack of resistance in the circuit.] Extreme corrosion at the ECT connector will lead to excessive resistance in the circuit. This condition will raise the ECT signal voltage leading the ECM into calculating that the engine is cooler than it is. A cold engine will make the ECM increase injector pulse-width, therefore, creating excessive CO at the tail pipe. In extreme high resistance conditions, engine flooding can occur, since the ECM reacts as if the engine temperature is at sub-zero temperatures. If on the other hand, a lack (shorted) of resistance is present at the ECT circuit, the opposite of the previous will hold true. Lower than normal ECT circuit resistance will lower the voltage signal across the sensor tricking the ECM to act if the temperature is higher than what it really is, which in turn makes the ECM reduce injector open time or pulse-width. Such a condition can cause a lean misfire or even a no start on a cold engine, due to the fuel starvation effect caused by the ECM's reduction of injector pulse width.

NOTE

Be aware of the fact that some computer systems, depending on the manufacturer, will substitute an erroneous ECT signal with an acceptable value on the scan tool. This is done so as to prevent the engine from stalling, allowing the driver to reach a repair shop. Always be mindful of the fact that what you see on a scan tool might be a substituted value and not the real thing. It should be standard procedure in all cases to back your diagnostic scan readings with an actual multi-meter or a hands-on measurement to double check your work.

In the event of an ECT sensor fault, the ECM usually looks at the IAT (Intake Temperature Sensor) as an indicator for engine temperature. **Also keep in mind that a low coolant condition will signal the wrong reading to the ECM. This can result in unnecessary ECT replacement.**

As a final note on conditions that affect the ECT sensor, it is worth mentioning a curious phenomenon. The ECT is also affected by the **general electrolytic conditions of the coolant fluid**. There are cases, due to lack of maintenance, where the actual coolant becomes slightly acidic. This turns the engine and coolant into a sort of battery, thereby, totally skewing the ECT sensor's reading. The slight acid content of the coolant will produce a small voltage, which can interfere with the sensor signal. This is partly the reason for the use of plastic ECT sensors by some manufacturers. The same acidic coolant can also create a thin film around the ECT's casing, preventing it from providing an accurate temperature reading.

COMPONENT TESTING

- Make sure that the cooling system is working properly. All cooling fans should operate as they should and the thermostat should not be faulty.
- The first thing to check for when the ECT is suspected is the actual coolant level. A low coolant level condition will make the ECT reading faulty, since there is no contact between the coolant and the sensor's thermistor.
- Make an overall inspection of the coolant condition. Verify that there is no voltage present at the radiator. Using a volt-meter, dip one probe into the coolant and the other to the engine body. No voltage should be present.
- Perform a visual inspection of the sensor and its connector. Check for sulfated or corroded connector pins that could cause a high resistance in the circuit.
- Perform a ground voltage drop to determine any faulty or loss of ground. Turn the key ON and simply connect the negative lead of the multimeter to battery negative and the positive to the ECT connector ground wire. With KOEO no more than 100 m Volts drop should be present and with KOEC no more than a sustained 300 m Volts voltage drop should be present.
- Verify the 5 volt reference line. Most systems are 5 volts. If in doubt check specifications.

- Connect a scan tool and monitor the ECT PID. At the same time connect a multimeter to each of the sensor's 2 wires (if it's a 3-wire sensor make sure you are probing on the sensor ground and the ECM's ECT signal wire). Get as close as possible to the ECM main connector so as to detect any wiring problem. Verify that the scanner ECT voltage PID reading is the same as the multimeter. This will rule out if the ECM is substituting the ECT value due to a problem or a possible ECM sensor ground fault that is skewing the ECT signal.

In some faulty systems, the vehicle's ECM will very briefly display the actual ECT sensor value on the scanner when the key is first turned on (a second or so). While turning the key on pay particular attention to the scan ECT reading.

- While monitoring both the **ECT sensor PID scan reading** and the **multimeter voltage** signal, disconnect the ECT connector and look for a reference voltage (most systems are 5 volt reference). When the connector is disconnected this will in effect create an infinite resistance across the ECT sensor circuit. An infinite (high) resistance translates into a very cold engine reading. Also wiggle the wires to uncover any intermittent wiring problems (short).

  Important! Do not expect an injector pulse width lengthening if the system is substituting the ECT value. This is precisely why the system substitutes a value so that the engine is affected as little as possible and allow the driver to reach a repair shop.

- Again while monitoring both the **ECT scan reading** and the **multimeter voltage** signal connect a jumper wire across the ECT connector and verify 0.00 volts (low resistance). This signals the ECM that engine is extremely hot. If both of these diagnostic steps pass the test, then it's an indication that the ECT wiring is good. Again also wiggle the wires to uncover any intermittent wiring problems.
- Disconnect the multimeter. While observing the scanner for a temperature change also take an infrared reading using an infrared gun. Compare both readings to detect if the ECT sensor is biased either high or low. Certain coolant additives and leak stops will coat the sensor with a film that prevents an accurate temperature reading. Although this is not an electrical problem per-se, it will cause an erroneous reading. This fools the ECM and skew the air fuel mixture. This final diagnostic step is intended to detect biased or shifted readings, which will not set off a faulty code.

This logical ECT sensor diagnostic routine will hopefully lead you in the right direction.

IAT SENSOR TROUBLESHOOTING STRATEGY

THEORY OF OPERATION

The intake air temperature sensor (IAT), also known as the air charge sensor(if screwed to the intake manifold runner), like the ECT is also a thermistor device. It measures the air temperature of the incoming air, and also like the ECT, it has a negative temperature coefficient. This means that as temperature increases its internal resistance decreases and vise-versa.

The IAT is a 2-wire sensor. One of the wires is an ECM supplied sensor ground and the other a reference voltage (usually 5.00 volts) that is also supplied by the ECM through a dropping resistor inside the ECM. As the air temperature changes, the IAT internal resistance also changes. This action also changes the voltage drop across the IAT sensor's thermistor.

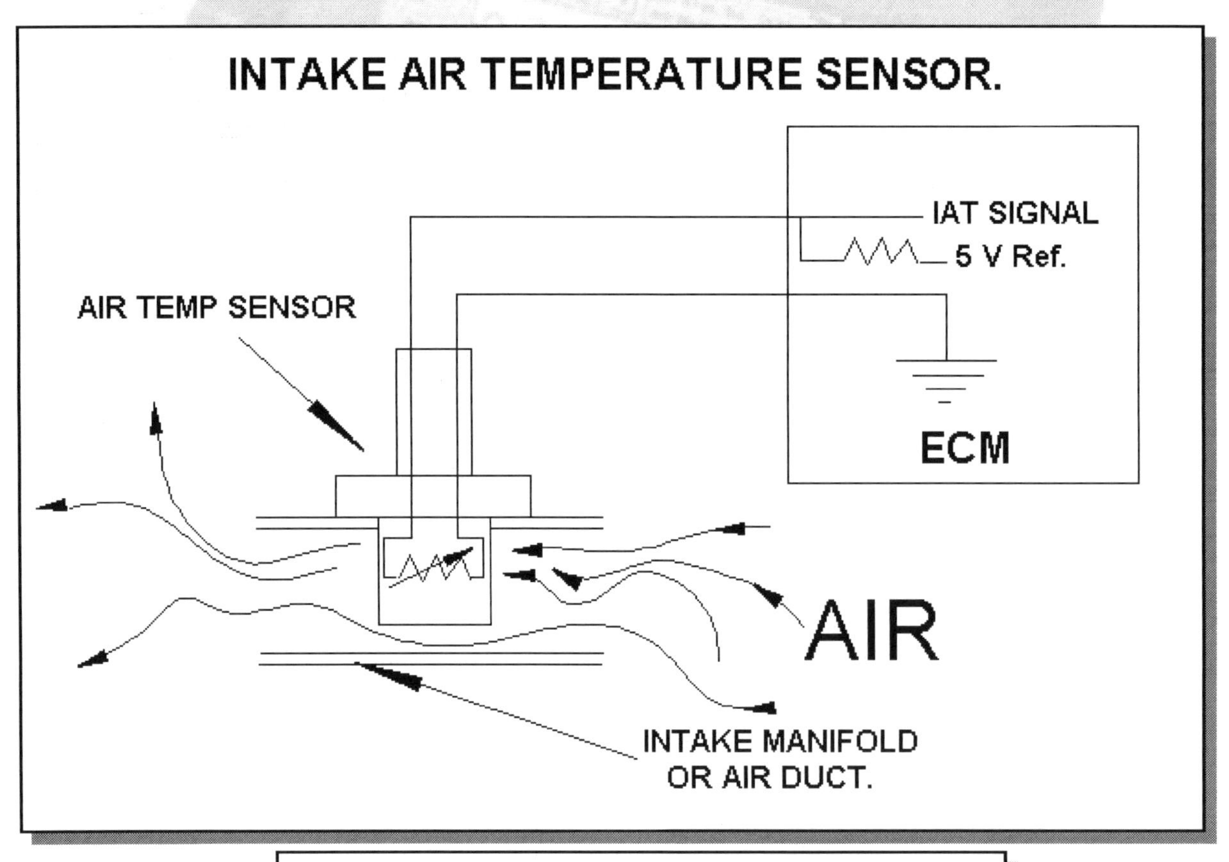

Fig 1 – IAT circuit and flow diagram.

The ECM uses the IAT signal to adjust the air-fuel mixture in accordance with air density. Air density changes with temperature. This means that a running engine on a cold day sucks in more air than on a hot day. Although the IAT measures air temperature, it is really telling the ECM how dense the air is. The ECM also uses this signal to modify spark advance, acceleration enrichment and determines when to enable EGR operation.

CONDITIONS THAT AFFECT OPERATION

The air temperature sensor is usually placed in the air duct, but in older vehicles it was also screwed on the intake manifold. The later was also called **"Air Charge Sensors"** and would get severely contaminated with carbon from fuel and oil buildup. Anything that prevents the IAT sensor from accurately reading the temperature of the passing air would send the wrong signal to the ECM. A rich mixture is usually the result of a contaminated or coated IAT, since the ECM would tend to react as if the temperature is colder than what it really is.

The IAT sensor can be removed and cleaned with carburetor cleaner, so long as it is not damaged. When this sensor is placed in the air cleaner, as in most new vehicles, it tends to last much longer. This placement shields the IAT sensor from the severe heat and carbon buildup inside the engine. Some of the import makes place it inside the VAF or MAF sensor, which provides a great deal of protection.

The ECM, as with the ECT sensor, may also substitute the IAT signal value in case of a fault. When reading the scanner's serial data, be aware of any signal substitution. An IAT scan reading that does not change or stays around 80 to 120 ° F could be a substituted value.

On late 70's injected Cadillac Sevilles, the ECM will have a hard time controlling fuel enrichment upon acceleration if the IAT sensor is defective. These systems will exhibit severe hesitation, since they depend greatly on this signal for fuel enrichment.

COMPONENT TESTING

The diagnostics testing routine for the IAT is similar to the ECT. Just remember that the IAT senses air temperature instead of coolant temperature.

KNOCK SENSOR TROUBLESHOOTING STRATEGY

THEORY OF OPERATION

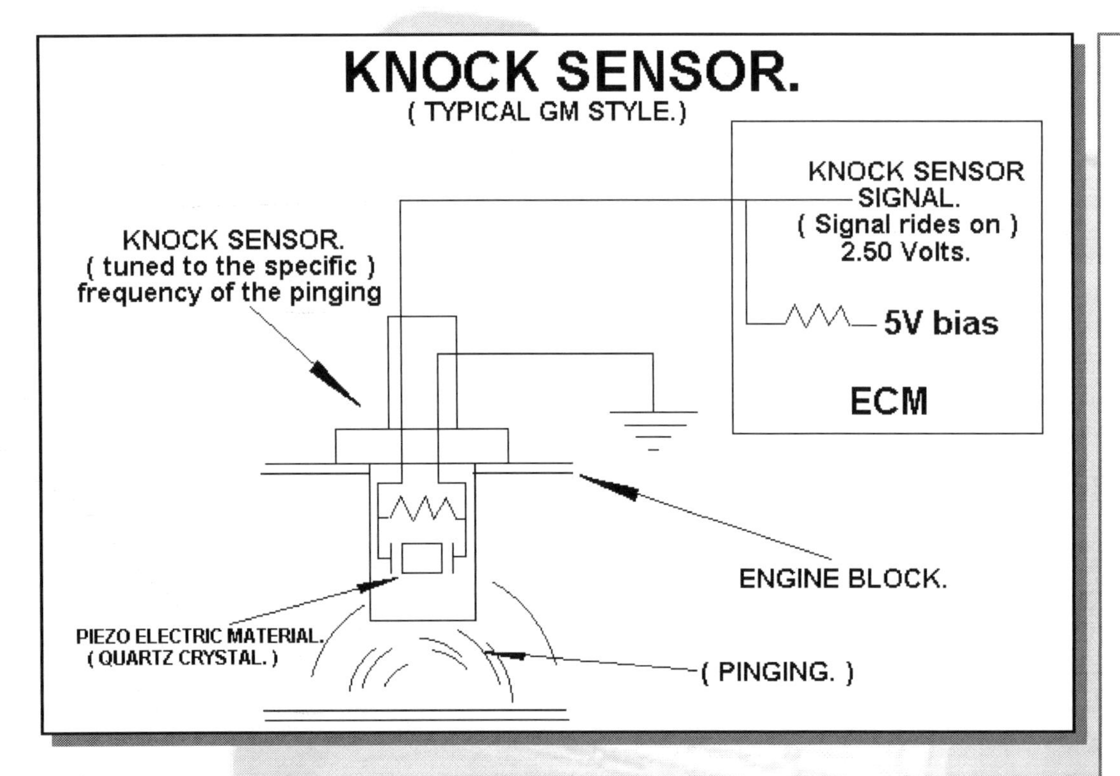

Fig 1 – GM knock sensor wiring diagram. The ECM provides a 5 volts bias to the knock sensor circuit. The sensor then divides the bias voltage in half, therefore, the output signal always rides on 2.5 volts. This is done to avoid ground noise interference, and also for diagnostics purposes, so that the ECM can detect an open or short circuit. This type of signal processing is called DC with an AC component.

In order to meet tougher emissions and efficiency standards, today's engines are burning less fuel than ever before. The government's CAFE standards (miles per gallon) or fuel efficiency has risen over the years to higher and higher levels. Because of higher efficiency, modern, leaner running engines also have to cope with the ever present pre-detonation or pinging problem. The leaner the air/fuel mixture the more prone it becomes to premature detonation. This is the actual explosion or combustion of the air/fuel mixture before the piston reaches TDC (top dead center) or the uppermost point of piston travel. Premature detonation causes the familiar pinging sound as if the engine's ignition timing was overly advanced. The actual engine components doing the pinging are the piston rings vibrating against the piston ring grove, although some engineers disagree on this subject. Such pinging can cause severe engine damage if left to itself.

The knock sensor was the answer to this problem. This sensor is actually a microphone screwed to the engine block, with usually only a single lead. The one lead (wire) sensor is grounded at the body. Two lead knock sensors are also used. The second version has the ground provided by the ECM. The knock sensor is responsible for detecting engine pinging. The sensor's microphone material uses the **piezoelectric effect** to do the actual knock detection. The piezoelectric effect states that when a crystal vibrates it produces an AC signal, as in this case when exposed to sound waves. The knock sensor is made of a crystal, **usually quartz**, and is tuned to the specific frequency of a pinging engine. In other words, the knock sensor only listens to a pinging engine and blocks all other sounds and noises that the engine can produce. This is true so long as the engine noise is not in the same frequency as what the knock sensor is tuned to.

The knock sensor is specifically designed for each particular engine and is not an interchangeable component. Special consideration has to be taken at the factory when tuning each knock sensor crystal element to the particular engine block, which the sensor is supposed to operate on. Each knock sensor is therefore unique in design.

Knock sensors in general are sometimes biased at some particular voltage level, usually 5 volts, although some manufacturers use different bias voltages. This means that the ECM provides a voltage on the signal line. This is done to avoid noise and interference associated with all ground circuits. The sensor itself divides the bias voltage in half because of its internal resistance. The knock signal therefore rides on the 2.5 volts bias voltage (if using a 5.00 V ref.). The knock circuit due to its higher voltage level (bias voltage) will not pick up any ground noise interference. The bias voltage also lets the ECM know when the knock sensor circuit has either open or short circuited.

NOTE

On some newer vehicles, the knock circuit has no particular bias voltage. In such cases, the ECM usually provides it's own filtered noise free ground to the sensor itself.

The knock sensor has a direct influence on the engine's ignition timing. The ECM uses the knock sensor signal to retard ignition timing and thereby reduce pinging. It is common to see a 2 degree or so ignition timing retardation in a step effect. This means that when the ECM sees a pinging engine it retards timing 2 degrees. If the pinging continues then it adds another 2 degrees of ignition timing retardation. The ECM keeps retarding timing in steps until the pinging stops. It is possible to see the effects of the knock sensor on engine timing by looking at a **graphed scan tool reading of the knock PID** and/or the **ignition timing PID.** By graphing these PID readings a relationship can be seen between the two signals. **Being able to graph the scanner's PID is of great importance in establishing relationships between two or more signals.**

PID stands for <u>parameter identification,</u> which means one single parameter or scan reading corresponding to a particular signal, sensor or calculated value.

CONDITIONS THAT AFFECT OPERATION

The knock sensor's sole job is to listen for engine pinging. Any type of interference with this microphone-like sensor will affect the ECM's ability to control timing.

NOTE

Most early European manufacturers connected the knock sensor to the ignition module (the EZL module) and not the ECM itself. This made sense to the Euro makers, since the knock sensor has a direct effect on ignition timing. Later models with more advanced computers (Motronic) integrated the knock sensor operation into the ECM itself.

Oil and dirt on the knock sensor's connector as well as electromagnetic interference from the ignition wires are all detrimental to the knock sensor operation.

Even though the knock sensor signal rides on a voltage bias, in extreme cases of deteriorated ignition wires, a spark arc could cause havoc in the knock sensor's signal line. This may cause the vehicle to exhibit lack of power, since the ECM is constantly retarding timing. The same goes for any mechanical problems present that may be causing an engine noise with the same frequency as an engine ping. A good example of this is a broken or cracked flywheel/flex plate. Such condition causes a noise similar in frequency to a pinging engine. The result is a noisy engine with severe lack of power due to the ECM severely retarding ignition timing. Always remember that on knock sensor equipped engines, a noisy mechanical fault can almost surely cause a lack of power symptom.

The sensor's crystal material itself could also get damaged, **making the knock sensor literally deaf**. Such a condition would render the sensor inoperative and in some cases the bias voltage would not be affected. If this happens, the ECM will not have the ability to detect a pinging engine and no possible ignition retarding would be available. Such an engine would ping severely without any action being taken by the ECM. On the other hand, modern OBD II systems will most likely set a code due to an inoperable knock sensor, even without a bias voltage. This is so because of the **functional nature of the OBD II system**. OBD II systems will always try to detect a faulty sensor before the fault actually happens. OBD II is sort of a pre-emptive system. The actual sensor does not have to be completely faulty for the system to set a code. It does this by running specific functional tests, called **"monitors"** during a **drive cycle**.

On older (EEC IV) FORD vehicles, at the end of performing a KOER test, you are instructed to perform a brief WOT. This is done in order for the ECM to test the knock sensor. When the WOT is performed, the ECM advances timing and listens for a knock signal which is supposed to hear. A lack of knock signal at this point will set a KOER code for the knock sensor.

COMPONENT TESTING

Testing the knock sensor is a simple mater. Simply probing on the signal wire with an oscilloscope and tapping on the engine block can generate a signal. Remember that it isn't necessary to tap too hard, since damage can be caused.

Set the scope to DC couple and check for the bias voltage first. ***If the knock sensor has two leads then it probably will not have a bias voltage.*** If a bias voltage is seen (usually 2.5 Volts), tap on the engine block with a suitable tool and check for an AC signal riding on the bias voltage line. On knock sensors with two leads (ECM provided ground) the actual AC signal rides on 0 volts and not on a DC bias voltage.

Remember not to set the scope on a very high time base. Knock sensor signals fall within the higher audio frequency mark in the frequency spectrum, which is quite low to begin with. If the scope is set too high, nothing will be seen at the screen.

MAF TROUBLESHOOTING STRATEGY

THEORY OF OPERATION

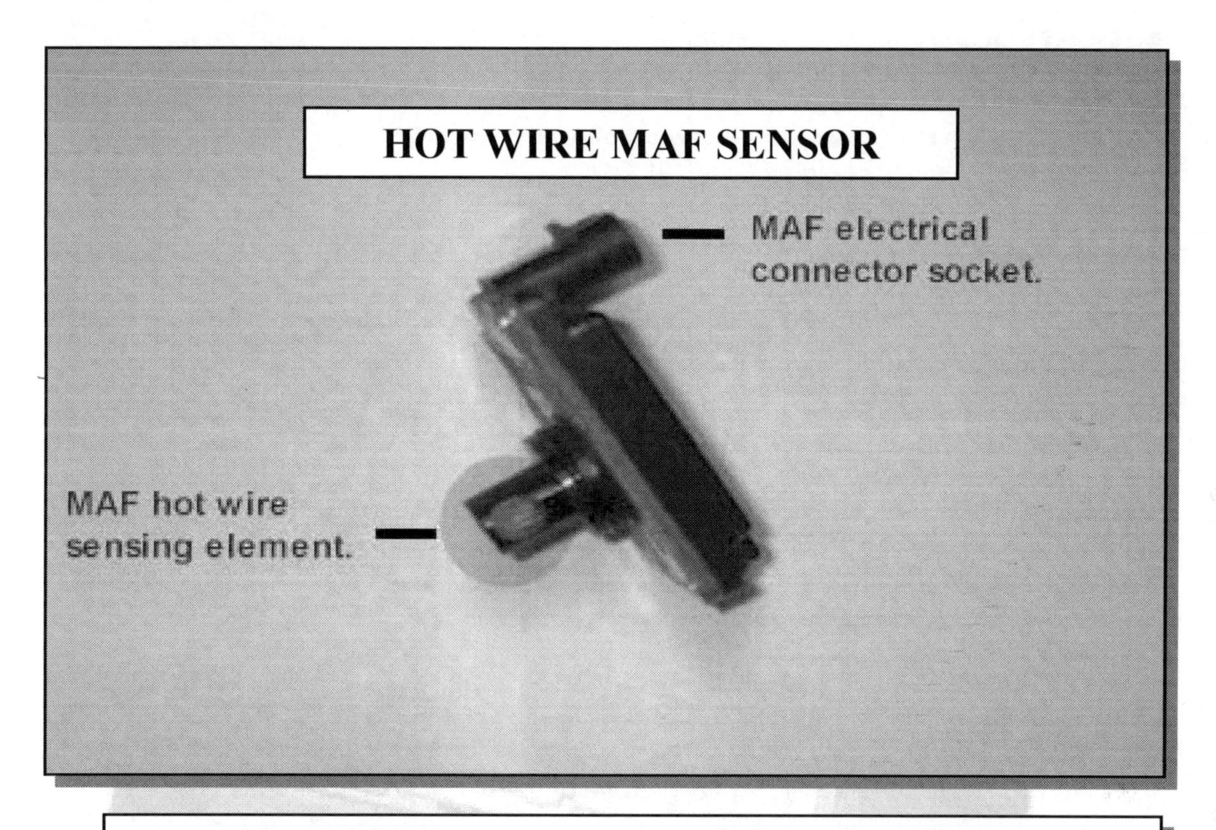

Fig 1 – Hot wire MAF sensor. Notice the fine sensing element.

The Mass Air Flow (MAF) sensor senses the amount of incoming air (Volume) into the engine. This sensor does not regulate the incoming air, this is done by the engine throttle plates. The MAF merely senses incoming air and relates a signal to the ECM. Air flow sensors come in three types. **The Vane Air Flow sensor**, **Hot Wire MAF** sensor and **Hot Film MAF sensor**. They all perform the same function but their operation is quite different.

- **The VAF sensor** measures the amount of air flow into the engine with a spring-loaded air flap/door attached to a variable resistor (potentiometer). VAF sensors measure air volume and not mass. The incoming air strikes or pushes against the internal air flap on the VAF sensor, which also moves the variable resistor's sensing arm (wiper arm). As air flows into the engine the mechanical air flap rotates further, causing the wiper arm to contact a series of resistors, changing the voltage signal output.

NOTE

The output signal on some systems is directly proportional (Motronic) to the incoming air, while in others the signal is inversely proportional (L-Jetronic). In other words, if the VAF electronic signal is directly proportional, then as the air flow increases the voltage output also increases and if the system is inversely proportional then as the air flow increases the output voltage signal decreases.
Directly proportional systems (Motronic) output a low voltage at idle with voltage increasing as air flow increases. Inversely proportional systems (L-Jetronic) output a high voltage at idle with the voltage decreasing as air flow increases. The ECM is programmed to the particular system and increases injector pulsation as more air enters the intake manifold.

- The **VAF sensor** has an air-fuel adjustment screw, which opens or closes a small air passage on the side of the VAF sensor. This screw controls the air-fuel mixture by letting a metered amount of air flow past the air flap, thereby, leaning or richening the mixture. By turning the screw clockwise the mixture is enriched and counterclockwise the mixture is leaned. In addition to the regular air flow measuring function, some VAF sensors also employ an air temperature sensor (IAT sensor) and a fuel pump switch.

  The **IAT sensor** is found inside the VAF casing and has the same electrical characteristics as a regular air temperature sensor. The VAF sensor flap also closes a set of contacts that activate the fuel pump relay coil (circuit opening relay). The contacts are actually closed as soon as the smallest amount of air pushes on the air flow flap. Once this happens the fuel pump starts running and the engine starts.

  One of the main drawbacks of the VAF sensor is that **it measures volume of air and not weight**. As air temperature changes so does its weight. There are more air molecules present when the air is colder than when it is hotter. As air temperature decreases, more air is absorbed by the engine, so there are drastic changes needed in the air fuel ratio (depending on the temperature of the air). The air temperature sensor inside the VAF somewhat compensates by signaling the ECM of any changes in air temperature.

- The **HOT WIRE MAF** sensor is a fully electronic unit. It senses the amount of air flow into the engine by measuring the amount of current needed to maintain a constant temperature through a very thin (70 micrometers) platinum hot wire. Hence the name hot wire MAF sensor. It also measures air by weight, since it takes into consideration the air temperature as well.

  This sensor works as follows. As the air enters the intake manifold through the hot wire MAF sensor it cools down the platinum wire, which is heated at a very precise temperature. When the MAF circuitry senses the platinum wire cooling down it increases the amount of current flow through the hot wire trying to maintain a specific temperature. This varying current flow is then converted to a voltage output signal by the MAF electronic circuitry and is used as an air flow indicator by the ECM. Hot wire MAF sensors have a signal that is directly proportional to air flow. So as air flow increases so does its voltage signal output.

  This sensor sometimes employs a mixture screw, but this screw is fully electronic and uses a variable resistor (potentiometer) instead of an air bypass screw. The screw needs more turns to achieve the desired results.
  A hot wire burn-off cleaning circuit is employed on some of these sensors. A burn-off relay applies a high current through the platinum hot wire after the vehicle is turned off for a second or so, thereby burning or vaporizing any contaminants that have stuck to the platinum hot wire element.

- The **HOT FILM MAF** sensor works somewhat similar to the hot wire MAF sensor, but instead it usually outputs a frequency signal. This sensor uses a hot film-grid instead of a hot wire. It is commonly found in late 80's early 90's fuel injected vehicles. The output frequency is directly proportional to the amount of air entering the engine. So as air flow increases so does frequency. These sensors tend to cause intermittent problems due to internal electrical failures. The use of an oscilloscope is strongly recommended to check the output frequency of these sensors. Frequency distortion is also common when the sensor starts to fail. Many technicians in the field use a tap test with very conclusive results. Not all HFM systems output a frequency. In some cases, this sensor works by outputting a regular varying voltage signal.

NOTE

A little known type of MAF sensor is the Karman-Vortex. Inside this sensor is an LED (light source), phototransistor, and a mirror (mounted on a spring base). The light beam from the LED is reflected back-and-fourth from the mirror and is picked up by the phototransistor. As the air flows though the sensor, turbulence or vortices are created inside it. These vortices cause the mirror to vibrate depending on the flow of air. It is these vibrations which are picked up by the phototransistor and converted into a frequency, which is sent to the ECM as a measure of air flow.

CONDITIONS THAT AFFECT OPERATION

VAF sensors are mechanical in nature. Their measuring element (wiper contact, pivot bushings and sensor resistors) get worn out over time. A binding air flap door is also a major problem with these sensors. The air flap mechanism is extremely precise and does not tolerate any misalignments. Always make sure that the air flap can travel freely all the way to its full open position. A broken air duct pipe will also render the VAF useless, since most of the air will be bypassed and enter though the broken duct hole. A thorough air duct check is always a good idea. The resistors also tend to wear out over time, sending the wrong voltage signal to the ECM. This will certainly throw off the air-fuel ratio.

NOTE

It is important to remember that most VAF systems do not have a potentiometer type TPS. These systems use a throttle switch sensor as an input for throttle position. Throttle switch sensors only signal the ECM of a closed or WOT throttle, but nothing in between. Therefore, the VAF sensor is the main indicator of throttle opening and load demand. Without the VAF signal input the engine will start and idle but will probably not be able to do much else, since the ECM does not know how much air is entering the intake manifold.

The air temperature sensor and the fuel pump switch are the other reasons for VAF failures. This fuel pump switch activates the fuel pump relay and its contacts also wear down over time, causing a no start-no no-fuel pressure condition. A simple continuity test will quickly reveal a bad fuel pump switch. The air temperature sensor also follows the same electrical characteristics of a normal IAT sensor and the same ohms to temperature tables could be used for diagnostics.

Hot Wire MAF sensors are very prone to sensing wire element contamination. A condition referred to by many technicians as **"growing hairs"** happens when debris, dirt from cheap air filters and outside air stick to the sensing wire element, shielding it from the incoming air. This shielding effect prevents the MAF sensor from correctly measuring the air flow and mass causing severe air-fuel ratio control problems. An ECM not in control while at pre-load is a strong indication of a dirty MAF.

In any fully electronic device, the electrical connections and circuitry fails after a certain lifespan of operation. An output signal voltage test will surely reveal a bad MAF sensor.

Hot Film MAF sensors tend to get electrical damage more often that the other type of sensors. The tap test ,as mentioned before, is a useful and simple procedure that usually reveals a bad hot film MAF sensor. Contamination or a broken air duct is also a problem for this sensors.

COMPONENT TESTING

- The first step common to all MAF or VAF sensors is to perform a thorough visual check of the air duct to detect any breaking of the rubber air duct. Secure and tight clamps are a must. Do not overlook this simple procedure as it is common to find broken ducts that are hard to see at first sight.
- The second common ailment to check for is a vacuum leak. Vacuum leaks have a big effect on MAF operation, since it provides a way for the air to enter the engine through a passage other than the throttle bore. This illegal air is never measured by the MAF and never compensated for by the ECM with extra fuel added. A lean mixture is usually the result. Check the power feed and grounds going to the air flow sensor.

VAF sensor

- Check for an air flap binding. The air flap should travel free through its entire travel range. Stick your fingers through the air inlet opening and push the air flap, sensing for any binding or mechanical problems.
- With KOEO the fuel pump contact should be checked by pushing the air flap very lightly and probing with a VOM on the fuel pump contact output pin to verify the output voltage. A failed continuity test of the fuel pump contacts is also a good failure indicator.
- With KOEO, verify the wiper arm output voltage (VAF output) with the air flap closed through the full open position. Compare readings to the correct specifications. Start the engine (KOER) and measure the output voltage at idle. Compare the measurement to correct specification. This test will detect a misadjusted air flow sensor. Remember that vacuum leaks or a broken air duct will derail the readings, making it seem like the air flow sensor is out of adjustment.
- Verify proper IAT sensor reading and compare to correct specs.

Hot Wire MAF sensor

- Check the voltage signal output KOEO and at idle.Compare to proper specifications. Off specs readings are common
- Accelerate the engine and look for a steadily raising voltage output.
- If the voltage output is wrong or no voltage at all, remove MAF and inspect the hot wire element. Look for a broken or contaminated hot wire element. Clean or replace as necessary. If MAF is dirty or contaminated check the operation of the burn-off circuit for correct operation. The burn-off relay should be activated for a second or so after the engine is shut off. Do not expect to see a RED HOT wire all the time. The hot wire element can be very hard to see, so use a multi-meter instead.

NOTE

A strong indicator of a dirty hot wire MAF sensor is the BARO scanner PID. The BARO or barometric sensor was eliminated between the late 1980's and early 1990's. Newer systems deduce the barometric pressure from the MAF sensor reading at WOT. The ECM simply assumes the correct barometric reading at WOT, since that is the time when the atmospheric and intake pressure are equalized. A dirty MAF sensor will be reflected on the BARO PID, making the ECM operate as if at higher altitudes. The results are disastrous to the A/F ratio control, with usually a lean condition as a result. Newer systems also update the BARO reading during brief acceleration periods.

In some systems, a quick check can be performed when the engine is not starting. Simply disconnect the MAF and start the engine. A non-starting engine that starts when the MAF is disconnected reveals a defective MAF unit. These systems will start on TPS alone and substitute the MAF reading with a good pre-programmed value.

HOT FILM MAF sensor

- Hot film units usually output a frequency. Check the frequency at KOEO and at idle. Compare to correct specifications. A KOEO off-calibration MAF sensor is most likely defective, since it already started out with the wrong frequency setting.
- Measure the output frequency throughout the entire RPM range. Be careful not to over-rev the engine.
- The outside casing of some of these sensors tend to melt or deform when defective. Perform a visual inspection and look for traces of melted material. In no other automotive sensor is the TAP test more important than with the HFM sensor. A missing of stalling engine when the HFM is tapped points to a defective unit.
- Using a scope, probe the output frequency and verify correct MAF sensor waveform integrity. A slight rounding off of the square waveform edges is sometimes normal. Experience and a good known waveform database are very useful.

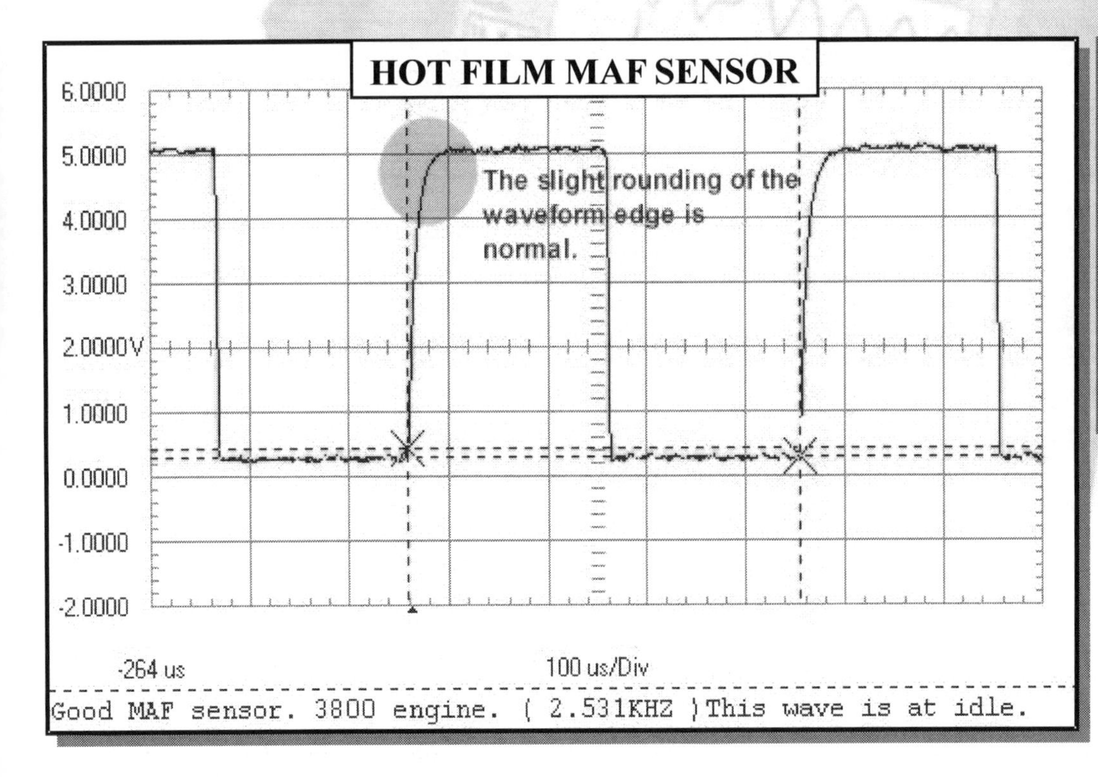

Fig 2 – Hot Film MAF sensor output signal at idle. Notice the normal slight rounding of the output square waveform.

These guidelines and a careful understanding of the particular system you are working on should point you in the right direction.

MAP SENSOR TROUBLESHOOTING

THEORY OF OPERATION

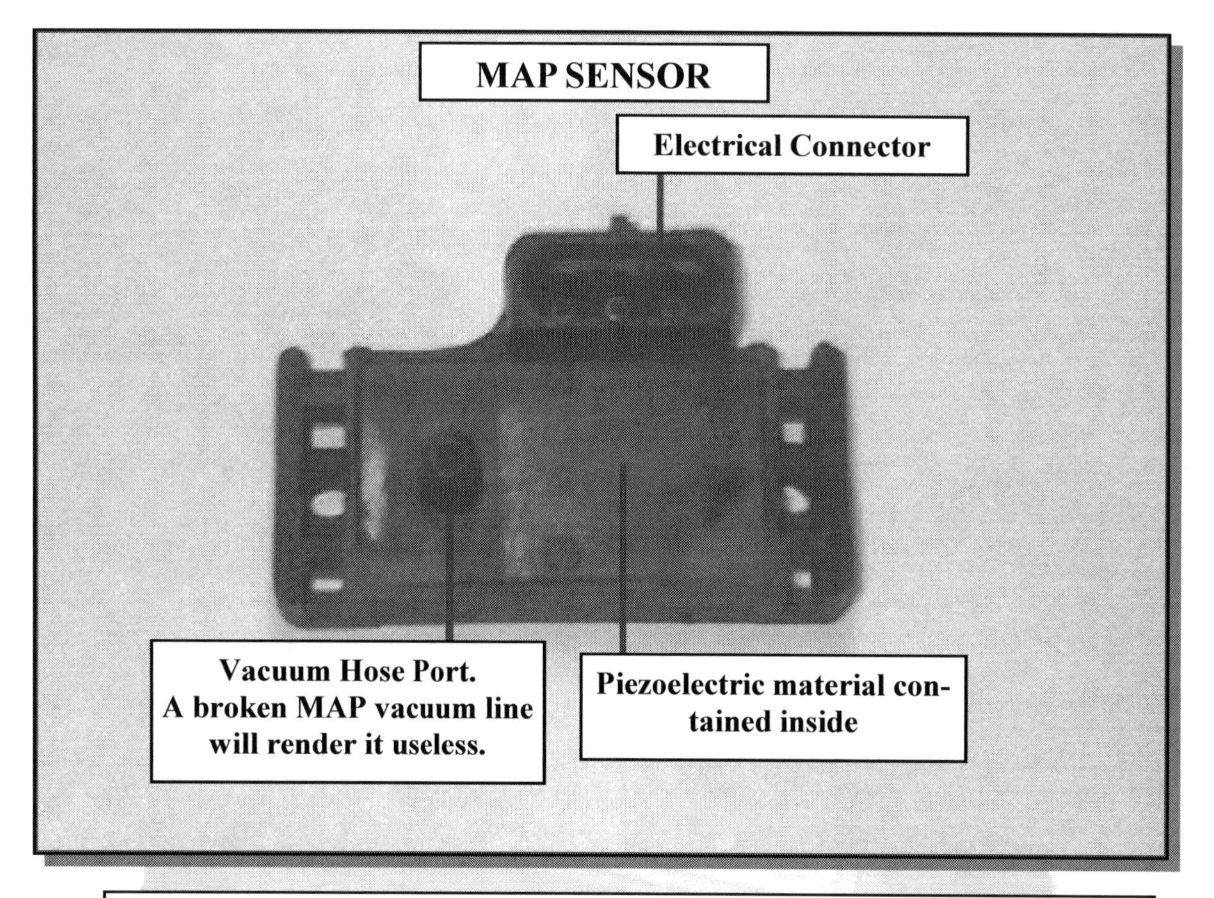

Fig 1 – GM MAP sensor. Notice the vacuum hose port on the left.

MAP sensors are three wire devices that measure intake manifold vacuum. In actuality, the MAP sensor measures the **difference** between intake manifold pressure/vacuum and atmospheric pressure. This is the reason why intake vacuum **is NOT the same** as MAP vacuum. Intake vacuum is atmospheric pressure minus MAP vacuum. With this in mind, the ECM makes the appropriate calculations as to the correct injector pulse. The MAP scanner PID is just MAP vacuum and should not be confused with intake vacuum. Few manufacturers do put out a manifold vacuum PID and Chrysler is one of them. In this case, the scanner PID for intake vacuum is a calculation (atmospheric press. minus MAP vacuum).

An engine's vacuum is a good indicator of load. The MAP sensor outputs a DC voltage or frequency and its signal is inversely proportional, which means that as manifold vacuum increases voltage or frequency decreases. MAP sensors are also used as barometric (BARO) sensors. As soon as the ignition key is turned on, the ECM reads the MAP voltage or frequency signal and automatically takes that reading as atmospheric pressure. Some manufacturers have configured their ECM programming so that the barometric reading is updated during a WOT condition. Once the engine starts, the ECM uses the MAP, TPS and RPM signals as main inputs to calculate fuel control on **MAP or SPEED DENSITY SYSTEMS** (systems without a MAF sensor). The ECM modifies injector pulse-width according to the MAP signal output or engine load. This sensor is also used for ignition timing and on some systems is a backup for the MAF sensor. With dual MAP and MAF systems, the MAP sensor is primarily used to monitor the EGR valve operation.

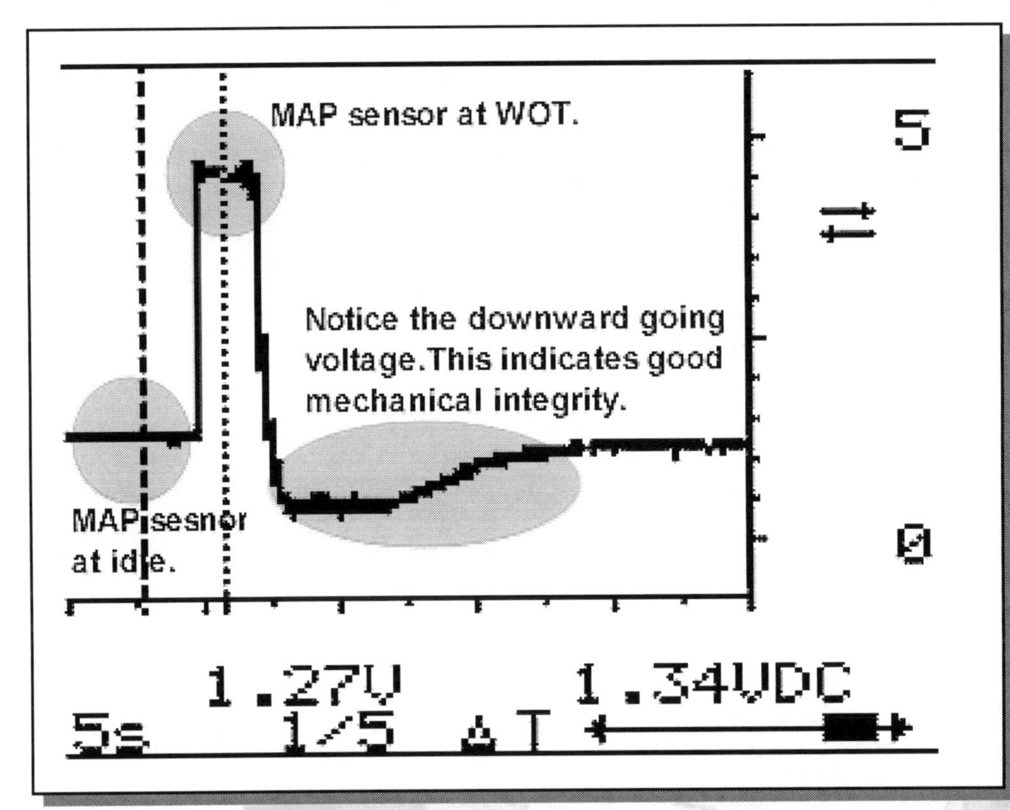

Fig 2 – Typical good MAP sensor signal. Notice the sudden raise in voltage after idle. This indicates good MAP response time.

	KOEO	KOER
GM	4.5 Volts at sea level to 3.2 Volts at 7000’	0.8 Volts to 1.4 Volts
FORD	159 Hz at sea level	95 to 112 Hz at idle
JEEP	4.6 Volts at sea level	1.1 to 1.3 Volts at idle
CHRYSLER	4.8 Volts at sea level	0.9 to 1.3 Volts at idle
TOYOTA	3.3 TO 3.8 Volts at sea level	1.2 TO 1.3 Volts at idle
HONDA	3.0 Volts at sea level	0.9 TO 1.1 Volts
VOLVO	4.6 TO 5.0 Volts at sea level	0.8 to 1.3 Volts at idle

Table 1—Typical readings for MAP sensors .

MAP sensors are made of a piezoelectric material. This material is a form of crystal (Quartz) that when bent changes its internal resistance. MAP sensors output two different types of signals. Most output a voltage and usually work with a 5.00 Volt REF. The other type found mostly on FORDs, output a square wave at a certain frequency (FORD uses 159 Hz). As manifold vacuum increases the frequency output decreases.

It is important to understand that the MAP and RPM or Distributor Reference signals are the two most important inputs to the ECM for fuel control. These signals should be quickly assessed and analyzed when encountering any fuel control related problems.

CONDITIONS THAT AFFECT OPERATION

MAP sensors are connected directly to manifold vacuum. This also means that any condition affecting the engine vacuum will also affect the MAP sensor reading. Conditions that affect engine vacuum are: **EGR stuck open, clogged catalytic converter, engine mechanical problems, vacuum leak, ignition timing problems, valve timing adjustments and low fuel pressure.** Also a shorted sensor feeding off the same sensor ground or 5.00 volt ref. line could cause a faulty MAP reading, due to the bad sensor shorting the MAP signal.

Beware of the fact that on a scan tool, some ECMs will substitute MAP reading on a rough running engine if it sees the signal out of range. Therefore the scan tool reading may be a substituted value that will throw the tech off during diagnostics.

COMPONENT TESTING

All MAP sensors have three electrical wires going to it: sensor ground, reference voltage, and signal wire. The sensor ground is provided by the ECM for all the other sensors. A ground voltage drop test should be performed, between sensor ground and battery post ground to verify no more that a 100 mV voltage drop during KOER.

The reference voltage is also provided by the ECM and is a 5.00 volts regulated feed line. It provides the MAP sensor with its working voltage. A shorted 5 volt reference line, either the line wire or another sensor that is shorting it will directly affect the MAP sensor reading and therefore the entire engine.

The signal line is the signal return to the ECM. It provides the ECM with the actual MAP sensor reading. This is the line to tap to when performing actual tests and comparing them to the tables found here. A shorted MAP signal line will also adversely affect the sensor's reading.

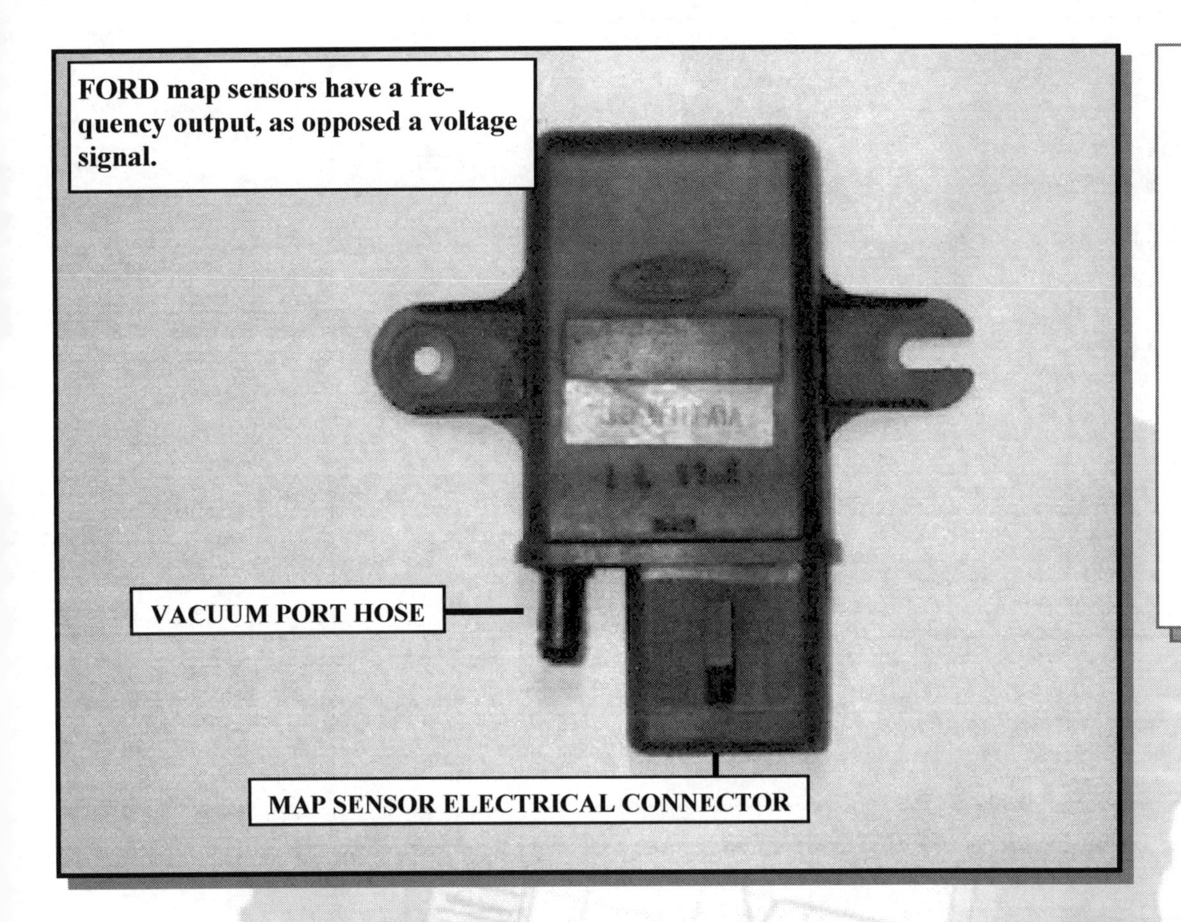

Fig 3 – FORD MAP sensors. At the lower right is the vacuum port. These MAP sensors have a frequency output and DO NOT put out a voltage signal.

- The first step in analyzing a MAP sensor is the KOEO signal reading. If this reading is wrong to begin with, the ECM will react as if the vehicle is operating at higher altitude and the engine will run richer, since we need more fuel at high altitudes.

- Second, make sure that the scan reading is the same as the actual reading coming from the sensor itself. This will verify that the ECM is not substituting the signal values.

- Third, make a careful analysis using the **"conditions that affect the MAP sensor"** to find any causes that may be inducing a problem.

- Last, there is always the possibility that the ECM is at fault. However, a defective ECM will almost surely be in the sensor ground or 5 volt reference line and will probably affect other sensors. This sequential diagnostic analysis will lead you in the right direction.

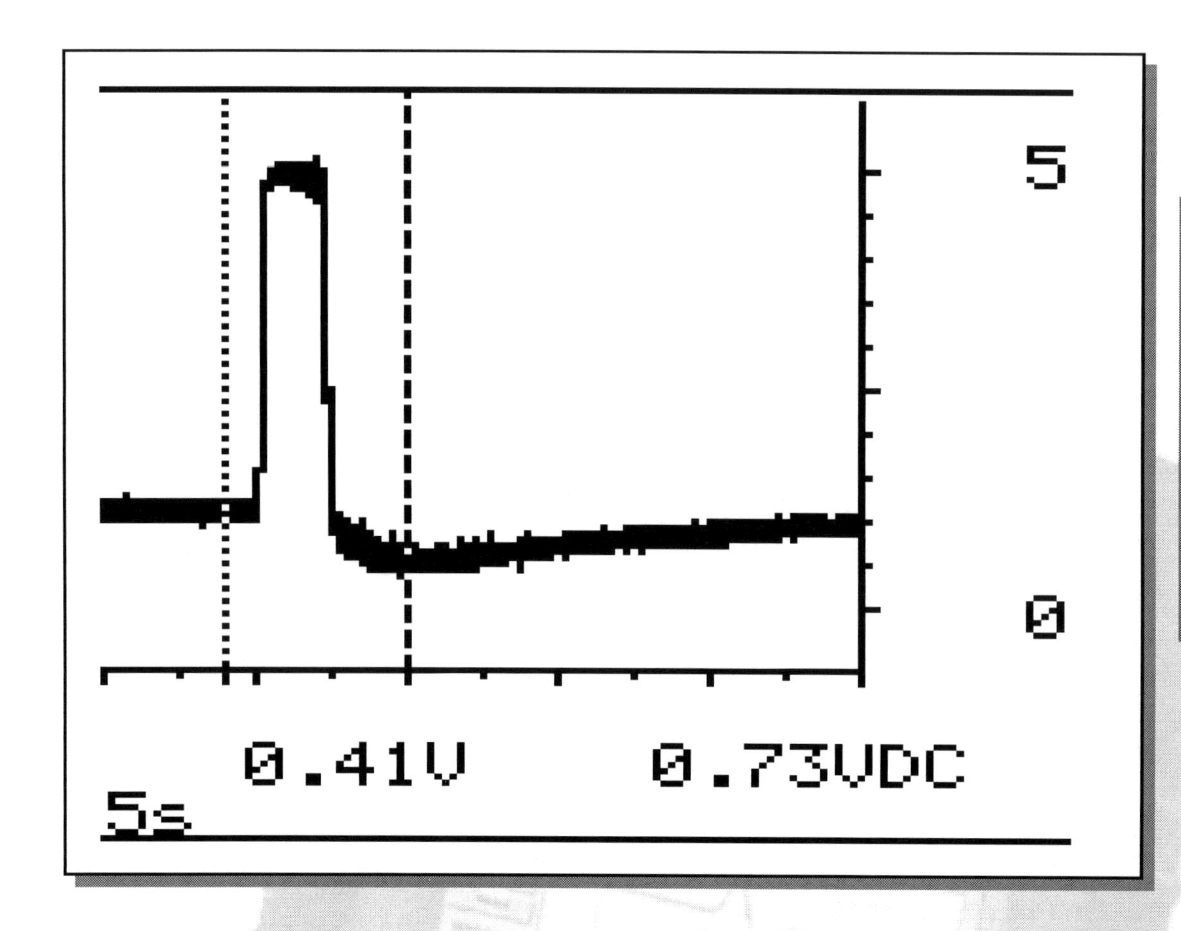

Fig 4 – Another MAP waveform from a good running vehicle. This signal belongs to a 93 Chevy Astro 4.3L eng.

By following these simple steps and carefully analyzing the corresponding wiring diagrams (s), it will surely point you in the right direction. If the reference voltage is gone, **DO NOT** assume that ECM is defective. If one of the sensors tied to the samc rcference line is shorted, it will bring down the reference line and all sensors connected to it. So, check your reference voltage.

TPS (THROTTLE POSSITION SENSOR) TROUBLESHOOTING

THEORY OF OPERATION

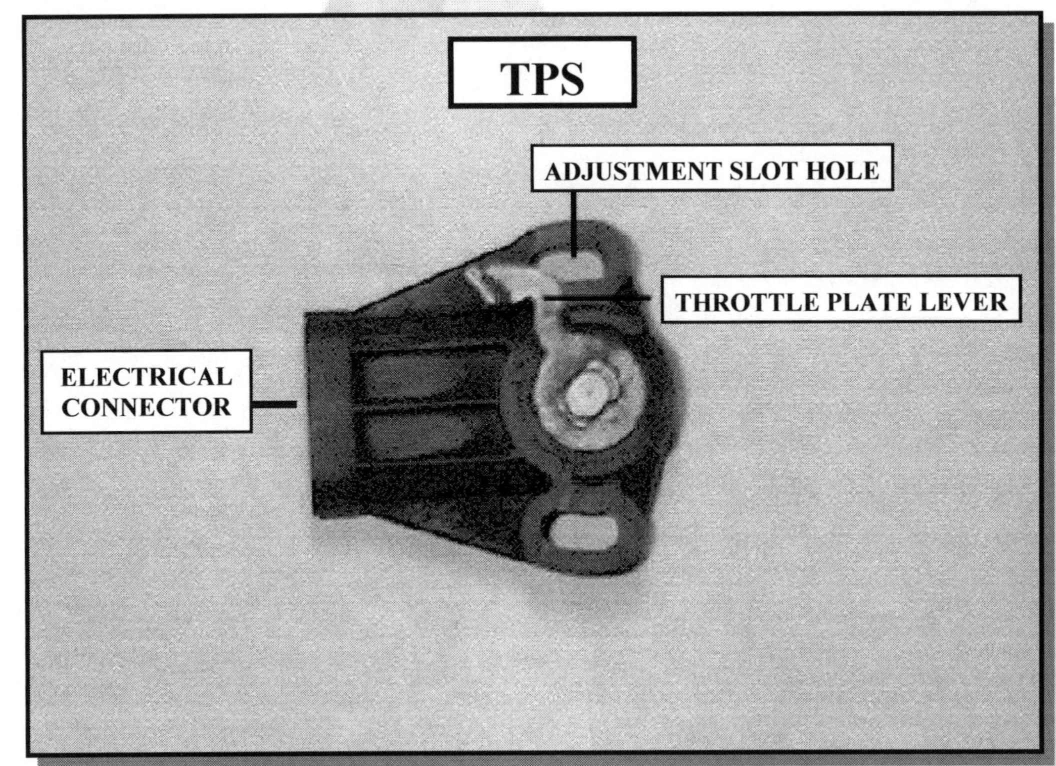

Fig 1 – GM throttle position sensor (TPS). Notice the slotted adjustment holes at each side of the TPS. These holes are used when performing a minimum air rate adjustment.

The TPS is a three wire sensor that measures the **throttle plate opening** as well as its **rate of change**. This sensor is a variable resistor, also called a **potentiometer**, that is directly linked to the throttle plate shaft. The TPS outputs a voltage directly proportional to the throttle opening. As the accelerator is depressed the throttle plate opens and the TPS voltage increases. This sensor is also one of the **main inputs to the transmission computer (TCM)**. The TCM uses the TPS input signal to control the transmission shift points and the torque converter (TCC) solenoid lock-up. The TPS together with MAP or MAF sensor are the main ECM indicators of acceleration and load. In other words, the ECM looks at these sensors to calculate engine operation upon acceleration.

Some manufacturers use the TPS signal as sole indicator of engine load when there is a faulty MAP/MAF sensor. In such cases, the MAP/MAF sensor values are calculated from the TPS signal. This means that the ECM substitutes the faulty MAP/MAF value from a look-up-table stored in its ROM memory, so the vehicle can continue to operate until the driver reaches a repair facility.

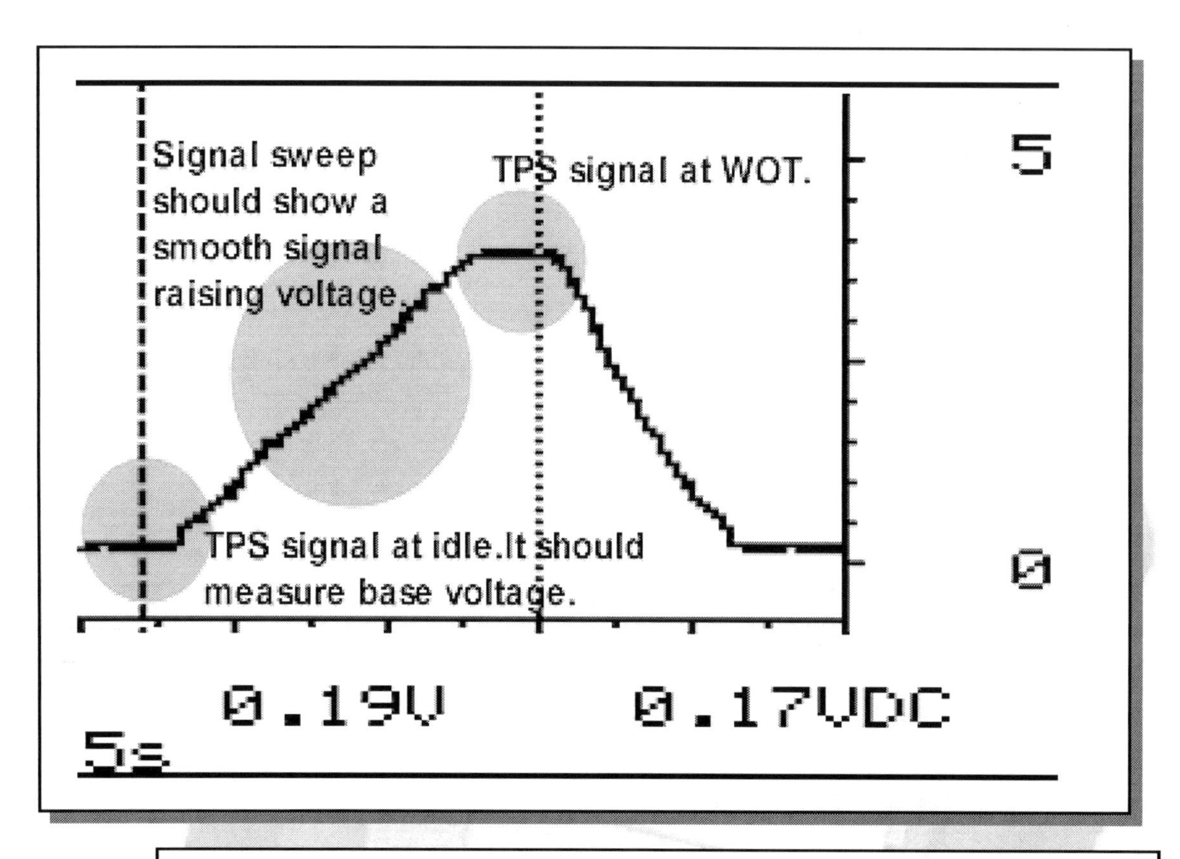

Fig - 2 – Typical TPS sweep waveform. Notice the smooth voltage signal change.

Typical readings for the TPS are:

KOEO		
GM	0.50 Volts	4.5 Volts WOT
CHRYSLER	0.60 Volts	3.5 – 4.5 Volts WOT
FORD	0.7 – 1.1 Volts	4.6 Volts WOT
JEEP	0.6 – 0.9 Volts	3.5 – 4.1 Volts WOT
HONDA/ACURA	0.5 Volts	4.8 Volts WOT
NISSAN	0.40 – 0.50 Volts	4.2 Volts WOT
TOYOTA/LEXUS (VTA signal)	0.50 Volts	4.2 Volts WOT
VOLVO	0.50 Volts	4.6 Volts WOT

The TPS sensors usually tends to fail at the lower range of its movement. This is where the driver is usually at, most of the time (as in cruising speed). This sensor usually works with a 5.00 volt reference voltage and ECM provided sensor ground. The signal is output through the signal wire, where all measurements should be made. Problems to any of the ground or power (5 volt) feed lines will cause an incorrect reading from the TPS.

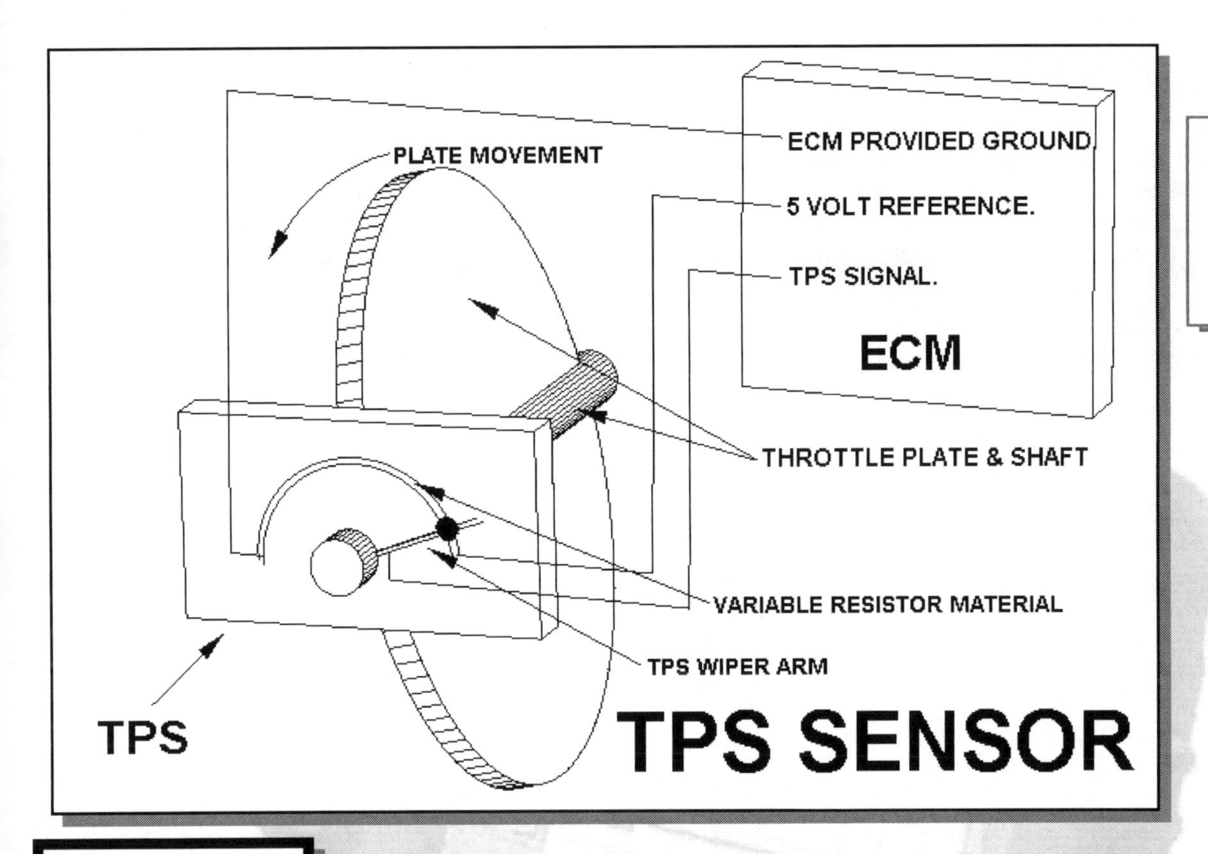

Fig 3 – TPS sensor & related wiring.

NOTE

TPS signal problems can greatly affect transmission operation. On a vehicle with an incorrectly shifting transmission, a careful analysis of the TPS signal should be made. A high TPS voltage reading during KOEO could also signal the ECM to go into FUEL CUT-OFF MODE. In which case, the ECM reacts as if the engine is at WOT and cuts pulsation to the injectors and preventing the engine from starting. Again, this is what is called fuel cut-off mode.

CONDITION THAT AFFECT OPERATION

Most TPS sensors **reset when the ignition key is cycled**. This means that whenever the vehicle is shut off and turned back on again, whatever voltage signal the ECM sees (TPS base voltage) it will take as 0 degrees of throttle opening. This ignition key resetting feature also means that as the throttle bore gets dirtier the idle speed/IAC operation will also be affected. Another curious drawback of a re-zeroing or resetting TPS is in the event of a **momentary signal drop-out glitch**. In this case, if the TPS signal momentarily drops to very low levels, the ECM will take the low reading as the 0.00 % or closed throttle point. Then, as soon as the output signal snaps back to normal, the ECM will perceive the new signal as a wider than normal throttle opening. The result is an increase in injector pulse-width and wrong transmission shift points. TPS sensors are directly linked to the throttle plate shaft. Any binding of the throttle plates can greatly influence the TPS signal. A carbonized throttle bore may also influence this signal. Some TPS sensors are combined with a throttle switch. This is a simple on-off switch that closes whenever the throttle is fully closed. This way the ECM knows that the throttle plates are closed. This combo TPS/Idle switch is mostly found on Euro and Asian imports. It is important to remember that a shorted sensor or actuator that shares the same power feed and ground with the TPS will also have an adverse effect on its signal. If a shared sensor shorts out it may also short the TPS reference voltage or ground line..

COMPONENT TESTING

TPS sensors are of the three wire type. The **sensor ground**, **reference (usually 5 volt) voltage**, and **the signal wire**. The **sensor ground** is provided by the ECM, as well as for all its other sensors. A voltage drop test should be performed across it and battery post ground to verify no more that 100 mV voltage drop during KOER. The **reference voltage** is also provided by the ECM and it is a 5.00 volts regulated feed line. It provides the TPS sensor with its working voltage. A shorted 5.00 volt reference line, either at the wire or another sensor that is shorted, will directly affect the TPS sensor reading and the entire engine. The **signal line** is the signal return to the ECM. It is the one providing the ECM with the actual TPS sensor reading. This is the line to tap when performing actual tests. A shorted TPS signal line will also skew its reading.

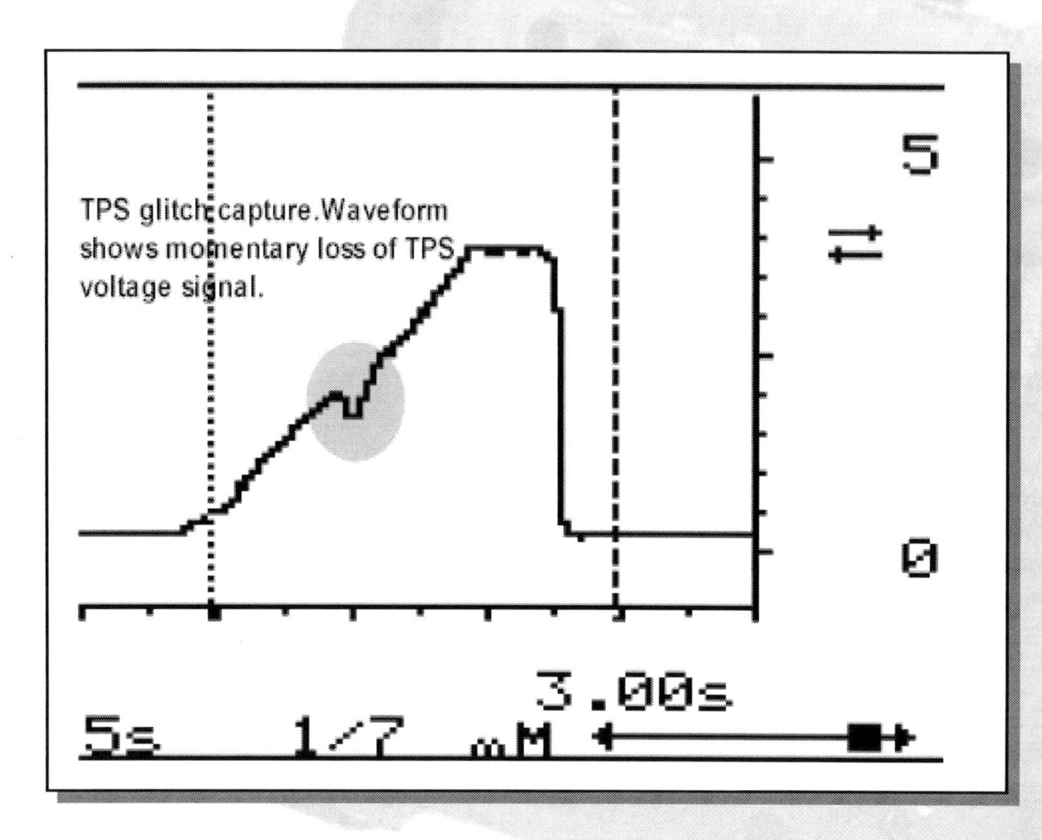

Fig 3 – Waveform showing a defective TPS signal glitch.

- The first analysis of the TPS signal voltage is done with the KOEO. This base voltage signal is taken by the ECM to be 0.00 % or degrees of throttle opening. It is very important that this signal be within exact specifications.

- Second, using a DSO sweep the TPS while measuring the voltage output. There should be no glitches or sudden drops in its voltage reading. ***Refer to Fig 3.*** Beware of the flat spots. Most TPS failures occur at the lower travel of the throttle plate. The TPS usually develops flat spots in this region and are not easy to detect. A faster frequency setting should be selected so as to be able to detect these flat spots.

- Third, make sure that the signal reaches close to the WOT voltage level. There is always the possibility that the throttle is binding or the TPS is defective at the high range of its travel. This would cause problems with vehicle hesitating at higher cruise speeds or upon heavy acceleration.

- Always make sure that proper idle speed is attained with the TPS fully closed. If idle speed stays high with TPS at base voltage the ECM will assume that it can't control idle speed and set a code.

NOTE

A minimum air-rate out of adjustment with TPS voltage too high or too low will, in many cases, cause an IAC (Idle Air Control) valve code. In such cases the scanner DESIRED idle speed can not be maintained because the throttle plates are over adjusted. When this happens, a reading of 0 steps on the IAC is seen on the scan tool and the idle speed will remain high. Properly adjusting the minimum air rate will correct this problem.

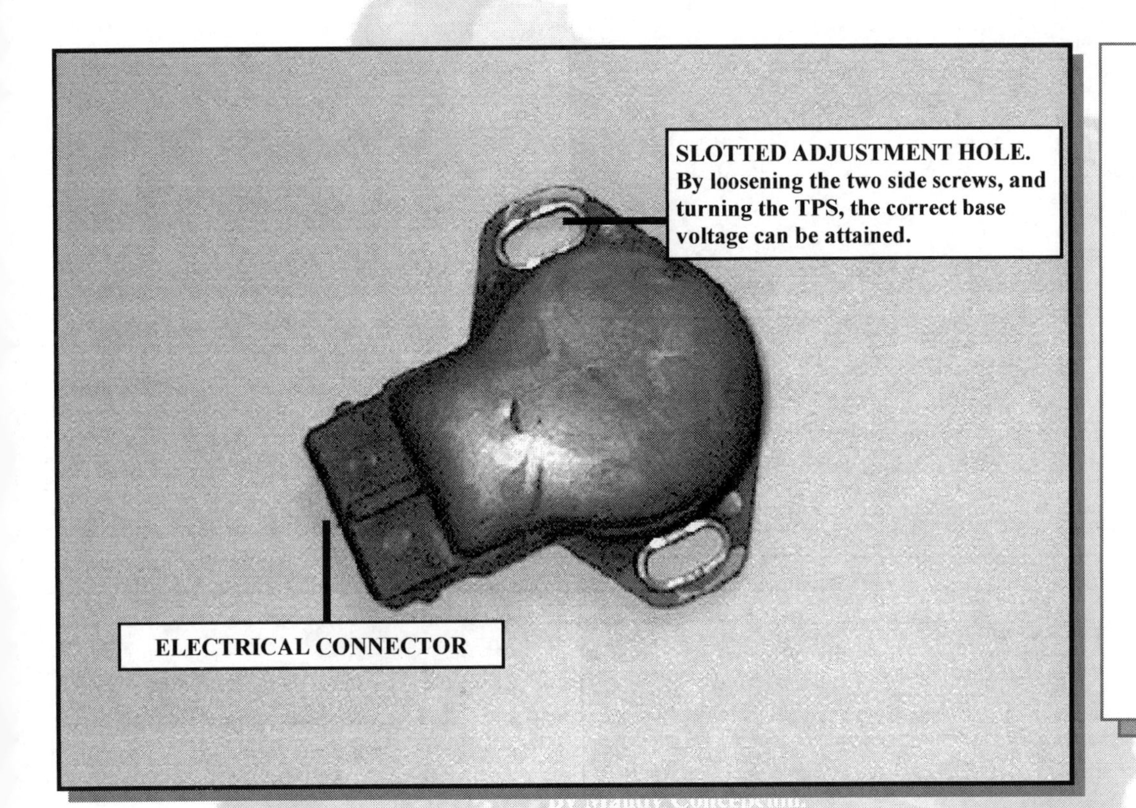

Fig 4 – MAZDA 626 throttle position sensor. This example also has the slotted adjustment holes at each side of the sensor. NOT all TPS sensors are adjustable, however. If it doesn't have the slotted holes it is not adjustable.

VEHICLE SPEED SENSOR/WHEEL SPEED SENSOR TROUBLESHOOTING STRATEGY

THEORY OF OPERATION

The **vehicle Speed Sensor (VSS)** has the job of providing the different modules, with **vehicle speed** and **deceleration factor**. This sensor is similar in operation to the CAM/CRK sensor and a couple of different variants are used. The VSS signal is used by the TCM (Trans. Control Module) to control shifting and TCC (Torque Converter) application; as well as by the instrument cluster module for speedometer operation. The ECM also uses the VSS signal to control fuel.

NOTE

Some manufacturers use the VSS, also called an OUTPUT speed sensor, to detect transmission slippage. In such cases an INPUT speed sensor signal is compared to the VSS (OUTPUT speed sensor) signal and a slippage factor determination is made by the TCM. These two signals are always being compared to an internal TCM memory table. If the signals are considered out of range, a trouble code is set and the TCM goes into limp-in mode.

The VSS may also be an input to the ABS control module. The ABS module uses the VSS signal to know the vehicle speed at all times as well as rate of deceleration.

There are a few types of VSS. These are **MAGNETIC, Magneto-Resistive, REED type and Photo-Electronic**. The MAGNETIC type for the most part is the most common one and it works in more or less the same fashion as the CRK sensor. It is always important to determine the type of sensor used. This will also determine the type of output signal that is to be expected. The magnetic sensor always puts out a **sine wave.** This type of VSS gets affected by anything that will decrease its signal amplitude, as in excessive air gap. The other VSS types put out a square wave. This makes the signal much more resistant to EMF, which is the reason why they are used. The reed VSS, for example, **has only 2 wires coming out of it.** This does not mean that it is a magnetic sensor, however. In this case a reference voltage is provided by one lead and a ground on the other. The reed VSS will simply ground this reference signal creating a square wave.

NOTE

Some manufacturers use a VSS with a built-in A/D converter to convert the magnetic sine wave signal into a digital square wave. An A/D (Analog to Digital Converter) is an electronic circuit that converts a sine wave into a square wave. In such cases where the A/D is built inside the VSS itself, the VSS also has to have a power and ground circuit. This is needed to drive the A/D circuitry. Some Toyota systems use this type of VSS.

The **wheel speed sensor (WSS),** on the other hand, is almost always of the MAGNETIC type. This type of sensor, as stated before, needs to have the right air gap to perform properly. However, newer late model systems (2004 & up) are starting to use hall-effect (square wave output) WSS. The reason for it higher resistance to EMF and less of a chance that the sensor may output a false reading.

CONDITIONS THAT AFFECT OPERATION

There are two main conditions that will greatly affect the performance of a magnetic WSS or VSS. One is the air gap between the sensor and the reluctor wheel **(also called the tone ring)** and the other is a shorted sensor coil. By far an improperly adjusted air gap represents a much more frequent problem. Dirt and oxidation will generally stick to the sensing part of the VSS/WSS and interfere with the air gap. An obstructed air gap translates to a faulty signal. On the other hand, a larger than normal air gap translates to a smaller amplitude waveform. **This presents a problem since most MODULES have a specific threshold recognition voltage, which is usually 1.00 volt P-P.** ***The respective module never recognizes a signal voltage that falls below the threshold recognition voltage level.***

CONPONENT TESTING

Testing the WSS is a fairly simple matter. With the right knowledge, a quick and accurate diagnostic is possible, even on hard to get places. These steps should be followed in the order presented here.

On the other hand, the VSS needs a slightly different approach to testing. This is because of the way the VSS signal reaches its applicable module. The VSS signal path should be traced to determine its operation. **This may also involve the trouble shooting of the "Data Bus Systems**. Therefore knowledge in data bus systems and how they work is also needed.

WSS TESTING.

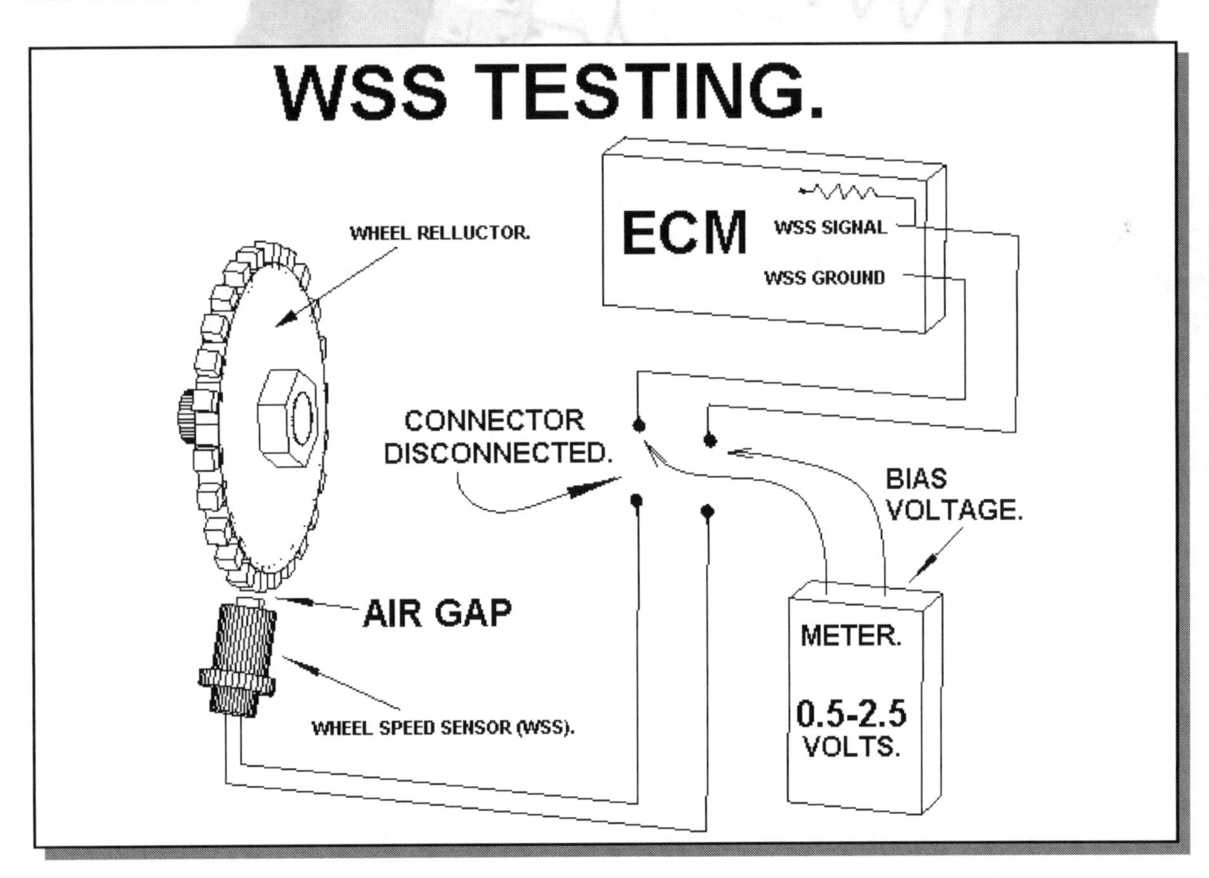

Fig 1 – Testing procedure for the WSS.

1. Scan the appropriate module and record any DTCs.
2. Using a scan tool, verify that the faulty WSS is not putting out a speed signal. A faulty sensor reading should be at 0.00 or 3.00 MPH/KPH without a signal output. If a scan tool is not available, then perform the tests manually.
3. Once the faulty WSS has been verified, proceed to perform a visual test. Follow the wires for the WSS and determine the location of a common connector. This will help in running further tests.
4. Once a common connector location has been found, proceed to verify for any **BIAS VOLTAGE. Note: The WSS is almost always a magnetic type. The applicable module sometimes puts out a bias voltage for diagnostics purposes. This voltage completes the circuit through the WSS coil and is used to detect open or short circuits. NOT all systems have a bias voltage, however.**
5. Compare the bias voltage of the faulty sensor to that of a good known sensor. If the faulty sensor harness voltage has 0.00 volts then there is an open or short to ground problem. A faulty sensor is the most probable cause. If the faulty sensor harness voltage has from 0.5 to 2.5 volts bias voltage, then the wiring is fine, go to ***STEP** 7*.

Some manufacturers DO NOT use a bias voltage. That's why it is always a good idea to check for bias voltage at a good known sensor first. Doing so will determine if the manufacturer is using a bias voltage for diagnostics.

6. If bias voltage is 0.00 volts at the faulty sensor's wiring connector and 0.5-2.5 volts at the module's wiring harness, then the open circuit is closer to the ABS module. Short the WSS harness connector using a jumper wire. Disconnect the main ABS connector and take an OHM reading. If close to 0.00 Ohms is seen then the wiring is fine. The problem is at the ABS connector.

Checking the WSS output signal.

7. Connect an oscilloscope to the two WSS output wires. While taking a scope reading spin the tire (at least once per second) and look for a uniform sine wave. The signal must be at least 1.00 volt P-P (Peak-to-Peak) to be considered good. A waveform with a small amplitude is an indication of an excessive WSS air gap or semi-shorted sensor coil windings. **Note: Most ABS modules in order to recognize a WSS signal need at least 1 volt P-P.**

VSS testing.

Although the WSS or the VSS sensor is usually of the magnetic type, it does differ greatly in the way the signal gets to the ECM or applicable module. The VSS is used extensively for transmission shifting in TCM applications as well as instrument cluster speedometer actuation. It is vital to determine its signal path before any diagnostics decision is to be made. This practice will speed up the diagnostic process. In this article various examples from different manufacturers will be shown in order to make the operation easier to understand.

1. As previously explained, trace the signal path using an electrical schematic diagram first. This will allow you to focus the final testing phase on the right component.
2. Determine all modules directly connected to the VSS. Once this is done, determine the condition of the sensor's wiring. Connect a scope to the sensor's signal lead at the specific module. While spinning the tires in the air, check for a VSS signal. If no signal is found proceed to a wiring check.
3. Disconnect the VSS and short the signal wire to ground. Using a VOM take a continuity check at the applicable module. If 0.00 Ohms is seen then the wiring is good, if NOT then an open circuit exists. In such cases proceed to check the VSS wiring for a breakage. Because of time constraints it might be desirable to simply run new wiring altogether.

In automotive electrical work, the use of a tone generator signal injector is very useful. Such inexpensive equipment is widely used by the phone companies to repair broken telephone wires.

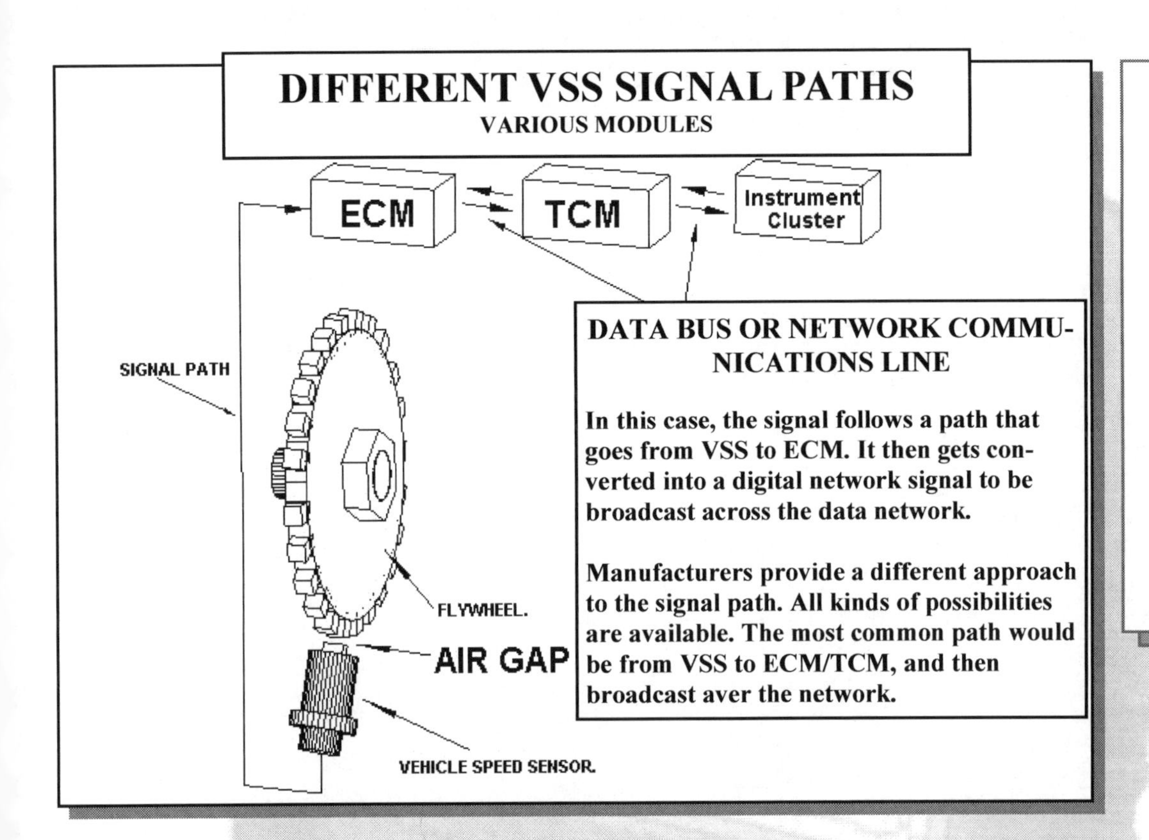

Fig 2 – VSS sensor signal paths. The different signal paths are shown. Not all-possible signal paths are referenced here and there are a number of other possibilities.

Example 1.

For example, Chrysler vehicles use a VSS-TCM-Data bus signal path on a number of their models. In this case, an inoperative speedometer could be due to the signal not reaching the data bus or that the data bus is down, since the **Instrument Cluster module** gets the VSS information from the data bus. For this reason, the Chrysler scan tool (DRB III) offers a menu choice PID for RPM reading at the data bus. If the RPM PID reading is seen at the data bus, then it is being transmitted. The fault is probably at the instrument cluster itself.

Example 2.

Most GM vehicles have a VSS-ECM-Class 2 data bus path. In this arrangement, the Instrument Cluster, Radio Control head, Chime controller and Cruise Control modules get the VSS signal over the data bus. Thc ECM is the only module that is hard-wired to the VSS and it is the one responsible for the data bus VSS signal transmission.
A no-VSS code on the instrument panel module, for example, right away points to a possible data bus or an instrument panel problem. This can be verified by simply scanning the ECM for a VSS signal while spinning the tires on a raised vehicle. If it does have an active VSS PID on the network and the other modules also see this signal over the data bus, then the instrument cluster or related circuit is at fault. By tracing the signal path we can determine if it is a data bus, wiring or a module problem.

Example 3.

Most mid-nineties FORD vehicles have the VSS signal hard-wired into all the applicable modules. In this case the VSS signal is not transmitted over the data bus. To correct any problems with this signal, normal electrical troubleshooting techniques are applicable. If a certain module is putting out a faulty VSS signal code then an OEM scan tool test is the preferred diagnostic choice. By simply locating the module with the missing VSS signal and proving the condition of its wiring, a quick determination can be made.

This type of wiring arrangement is fast disappearing, however. Such a system with all the sensors hard-wired makes use of excessive copper wiring. This adds a lot of weight to the vehicle with a definite cost in fuel efficiency, which is the reason why most manufacturers are going the data bus way. By transmitting as much data as possible over the 1 or 2 wire data bus, a massive amount of wiring can be saved. This translates into a fuel-efficient and simpler to repair vehicle system. **The use of different data bus schemes like Class 2, UART, SCP and CAN will be addressed elsewhere is this book.**

A/C PRESSURE SENSOR

THEORY OF OPERATION

In recent years, the heating and air conditioning system on automobiles has also been greatly affected by advances in automotive electronics. Starting in the late 1980s, new electronic HVAC (Heating Ventilation and Air Conditioning) systems started to appear in the more expensive luxury vehicles. The new HVAC systems made use of all kinds of different modules, sensors and solenoids to control the refrigerant pressure as well as the flow of air through the system. The new systems came to be called **"Climate Control Systems",** and as in any complex electronic system, technicians needed the appropriate knowledge and skills to diagnose them.

The early A/C pressure sensors were not really sensors at all, but simple ON/OFF switches. As technology advanced, the A/C pressure sensor became a full-fledged variable resistor. The A/C pressure sensor, also called the A/C pressure transducer, is a piezoelectric (crystal) device. A small crystal component inside the sensors casing, usually quartz, provides a varying resistance according to the sensed pressure. Over the years these sensors have gotten much more reliable and today are found in almost every climate control system on the market.

The main purpose of this sensor is to provide a specific engine load signal to the ECM. In other words, the sensor simply relays a pressure signal back to the ECM that is directly proportional to the refrigerant pressure. From this signal, the ECM infers the appropriate load factor according to a pre-programmed memory chart. If the A/C pressure sensor's working voltage is from 0.00 volts to 5.00 volts, the ECM already knows the relationship of voltage-to-pressure. Once the ECM calculates the A/C line pressure, a load factor can be determined. The ECM then uses this factor value to adjust timing as well as injector pulse duration. Any particular problem with this sensor can cause severe excess emissions as well as engine performance problems, given that the ECM can not calculate the A/C load factor.

The A/C pressure sensor is a three-wire type (the old A/C switch has two wires). The ECM usually provides a ground as well a reference voltage (usually 5.00 to 8.00 volts). The sensor then sends a modified signal back to the ECM, which will normally range from 0.00 to ref. voltage according to the line pressure. A shorted sensor (0.00 volts) will signal the ECM that the A/C is at maximum pressure. The ECM will interpret this as a load on the engine, although not a real one. In this situation, the ECM will change the entire engine operation to meet the higher load demand. The result is a wider injector pulse, shifted ignition timing as well as higher idle air control command, which will greatly affect performance. One of the first things to look for, whenever diagnosing erratic idle control problems, is the A/C pressure switch PID reading on the scanner. It is also important to know that this sensor's reference voltage and/or ground is sometimes shared with other sensors as well. In such cases, if the pressure sensor shorts out it could also render the other sensor (s) that share its reference voltage or grounds inoperative. Such is the case as with some Chrysler vehicles where the A/C pressure sensor reference voltage is also shared with the MAP sensor. **A shorted A/C pressure sensor will cause the vehicle not to start due to an inoperative MAP sensor.** Conditions like this one require a careful analysis of the particular wiring diagram. Once a relationship is determined by looking at the shared reference and/or ground wires, a conclusion can be reached. By simply monitoring the circuit and disconnecting each shared sensor, the shorted component can be found. The important thing is to realize which sensors shares the same wiring.

The sensor's signal path varies with each manufacturer. The sensor is usually connected to the ECM where it then controls the various A/C components based on the A/C pressure sensor signal input. The A/C pressure sensor's signal can also be shared with other modules such as the ATC (automatic temp control) or the BCM (body control module) to then manage the different HVAC actuators and relays. But regardless of the signal path it always performs the same function, which is to provide the appropriate module with the necessary data to control engine and HVAC operation.

CONDITIONS THAT AFFECT OPERATION

A shorted A/C pressure sensor, as mentioned before, will cause severe engine performance problems including excessive emissions and faulty idle operation. This signal is basically a load input to the ECM, since it is an indication of A/C compressor operation. Any electrical problems with the pressure sensor's signal can also influence other sensors that share the same ground and/or reference voltage. A shorted sensor, will cause any other important sensor (i.e. MAP sensor) that shares the same wires with it, to also fail. For this reason, it is possible for an A/C pressure sensor to cause a no-start condition.

The A/C pressure signal can also be affected by a low refrigerant condition, which will cause the signal not to change when the A/C compressor is activated. A low refrigerant condition may cause many automatic temperature control modules to go into a faulty mode and deactivate the A/C compressor, in which case, a scan tool should be used to erase and re-set the stored codes after recharging the system. Failure to do so will cause the A/C system not to turn on, even if the low refrigerant condition is corrected.

COMPONENT TESTING

The testing of the A/C pressure sensor centers on the three leads that connect to it. The following procedure is divided into two parts, which are the sensor itself and the wires that connect to it.

WIRES

- Make sure that the A/C system is fully charged with refrigerant.
- Disconnect the A/C pressure switch.
- Using a voltmeter measure the voltage between the reference wire and battery ground. The meter should read reference voltage. This will test the reference voltage (usually 5.00 volts).
- With a voltmeter measure the voltage between the sensor's ground wire and battery voltage. The meter should read battery voltage (checks for proper sensor ground).
- With a voltmeter measure voltage between the sensor's connector ground and reference voltage wire. The meter should read reference voltage (double-checks and assures proper ref. voltage and ground wires).
- Jump the signal wire to ground and then to the reference voltage at the connector. Using a scan tool, read the A/C pressure sensor reading. The scan tool should read 0.00 and reference voltage respectively (tests for signal wire integrity).

It may be necessary to either switch the ignition key on or start the vehicle (turn A/C on) to perform the last part of this procedure. Some systems will not output a data stream if the engine is not on.

SENSOR

If all these tests checked out OK, then the problem is probably at the A/C pressure sensor itself. Recover the refrigerant if necessary and remove the A/C pressure sensor from the refrigerant line. Reconnect the sensor and using compressed shop air, apply air pressure to the pressure port. As pressure is applied and released a varying voltage should be observed at the signal wire using a voltmeter or the scan tool. If not then the sensor is at fault. The idea is to isolate the faulty circuit or sensor in order to arrive at the final diagnostic conclusion. This will prevent unnecessary replacement of parts.

FUEL RAIL PRESSURE SENSOR (FRP) STRATEGY

THEORY OF OPERATION

In recent years the return-less fuel system has gained wide acceptance. With this type of system, the return fuel hose is eliminated in favor of a different type of fuel pump module, using a built-in fuel pressure regulator. In an in tank return-less fuel pump module system, the fuel is returned to the fuel tank right at the fuel pump itself without ever leaving the fuel tank. The reason for all this is to prevent an unnecessary amount of fuel from reaching the engine bay, where it will heat-up and cause excessive fuel vapors at the fuel tank. In other words, this system addresses the never-ending struggle to stop excessive EVAP emissions.

Another high-tech approach to the fuel vapor problem is the use of a fuel rail pressure sensor in conjunction with a variable-speed electric fuel pump. Ford, among others, has adopted this approach in a wide variety of their newer model vehicles. With the electronic return-less system, the ECM relies on the FRP (fuel rail pressure) sensor for fuel pressure input right at the fuel injectors. By monitoring the fuel pressure, the ECM can then adjust the fuel pump's rotational speed and maintain a stable pressure. Once a stable fuel pressure is attained, the formation of fuel vapors in the fuel line itself is greatly reduced. The whole process happens very fast since it is electronically controlled.

The FRP sensor is a three-wire piezoelectric electronic pressure sensor. This means that the sensor's resistance varies as pressure changes. The FRP sensor is also connected in line with an internal ECM voltage divider resistor network. So that as the sensor's resistance changes with pressure the overall current flow varies as well. The higher the sensor's resistance the less current flow and the higher the voltage. The higher voltage across the FRP sensor will cause a lower overall voltage across the ECM's internal resistor and vise-versa. A typical FRP sensor voltage-to-pressure chart is shown next.

FRP sensor voltage chart.

0.50 volts – 0 PSI
1.2 volts – 10 PSI
1.65 volts – 20 PSI
2.2 volts – 30 PSI
2.75 volts – 40 PSI
3.45 volts – 50 PSI
3.9 volts – 60 PSI
4.6 volts – 70 PSI

Fig 1— Typical fuel rail pressure (FRP) sensor voltage chart.

In some FRP sensor applications, the sensor is also connected to the intake manifold side. In this arrangement the sensor's signal output is a differential signal of fuel pressure to intake manifold, which the ECM uses to control the fuel pump speed. Therefore, maintaining the fuel in the rail in a liquid state and preventing fuel vapors.

NOTE

The differential signal of the FRP sensor takes into consideration the amount of intake manifold vacuum of the engine. This way the ECM can properly control the actual amount of fuel leaving the injectors. In other words, the ECM actually tailors the fuel delivery according to the engine operating demands. It takes less fuel pressure to push a certain amount of fuel through an injector connected to a high vacuum manifold then otherwise. An engine having high vacuum creates a suction at the injector manifold ports, and the result is less pressure needed to inject the fuel.

CONDITIONS THAT AFFECT OPERATION

The FRP sensor is connected directly to the injector fuel rail, which makes it susceptible to the same temperature variations as the injectors. A clogged fuel filter, a defective in-tank fuel pump module, dirty fuel lines, etc could also cause an erroneous signal reading. It is important to determine if the faulty signal reading is caused by the FRP sensor itself or some condition that is affecting it. Do not condemn the FRP sensor until all the necessary testing have been performed.

COMPONENT TESTING

The FRP sensor is a three-wire type sensor. The ECM provides a reference voltage as well as a signal ground to the sensor. The sensor then sends a pressure signal back to the ECM thought the signal wire. It is also good to know that the FRP sensor may shares the reference and ground wires with other sensors and any electrical conditions that affect the other shared sensors will also affect it. Follow the steps bellow to determine the root of the problem.

- First determine if there is actual fuel pressure in the system. This will eliminate a faulty reading condition caused by a mechanical fuel system problem. Using a fuel **pressure/volume gauge**, ascertain that the system is working properly.
- Disconnect the FRP sensor and open the ignition switch.
- Using a voltmeter, probe between the FRP reference voltage wire and battery ground. Reference voltage (usually 5.00 volts) should be seen at the meter (tests for proper reference voltage).
- With a voltmeter, measure the voltage across the FRP sensor ground wire and battery positive. Battery voltage should be seen (tests the integrity of the FRP sensor ground circuit).
- With a voltmeter, measure the voltage across the reference and ground wires of the FRP sensor (double-checks the reference voltage and ground leads of the FRP sensor).
- Jump the FRP sensor signal wire to the ground wire at the connector. Using a scan tool, monitor the fuel rail pressure PID (with engine off). About 0.00 volts should be seen.
- Jump the FRP sensor signal wire to the reference voltage wire at the connector. Using a scan tool monitor the fuel rail pressure PID (with engine off). A reference voltage reading should be seen (usually 5.00 volts).
- Make certain that the vacuum hose going to the FRP sensor is not clogged or broken. The engine vacuum is taken into consideration by the ECM when adjusting the actual fuel pressure. The FRP sensor will actually output a differential signal, which takes into account the amount of vacuum at the tip of the injectors.

If these steps check out OK and there is no fuel system mechanical problems, the fault is probably at the FRP sensor itself. Take extreme caution when replacing the FRP sensor, since you will be dealing with flammable fuel. **Always be aware of the fire extinguisher's location** and avoid any open flames while working on the fuel rail. Follow the manufacturer's replacement procedures.

PFS (EVAP PURGE FLOW SENSOR) DIAGNOSTIC STRATEGY

THEORY OF OPERATION

The PFS (EVAP purge Flow Sensor) is a little known device used in some Ford vehicles to detect the EVAP purge valve operation. The ECM uses this sensor during the EVAP monitor execution. In essence the PFS is like a mini MAF sensor in that it detects the flow of a gas. In this case it detects the fuel vapors flowing into the intake manifold.

There are two variants to this sensor, the two or the three wire type. The two wire PFS sensor is a simple thermistor that changes resistance with the flow of gas, such as the EVAP gasses flowing during purge. This thermistor works the same way as any other, also with a negative temperature coefficient. The three-wire PFS sensor differs somewhat from other 3-wire sensors in that it uses battery voltage as its reference instead of the normal 5.00 volts used by most sensors. The voltage supply wire to the PFS is provided by the main ECM relay. The other two wires are the ground and signal wires. The PFS however does not send a 12 volts signal back to the ECM. It converts the battery voltage and sends a varying voltage back to the ECM according to the amount of EVAP flow detected by it. During normal operation the PFS signal climbs to between 3.7 and 4.6 volts when the ignition key is first turned on and about two seconds letter it ramps down to between 1.5 and 1.9 volts. If the PFS signal voltage stays high or low all the time, suspect a faulty sensor. These sensors have a high failure rate. As the ECM commands the purge valve on, the voltage should increase slowly from the baseline of 1.5 volts steadily. A defective PFS will usually cause a P1443 code to set in memory.

COMPONENT TESTING

The following procedures should be used when testing the 3-wire PFS. The 2-wire (thermistor type) PFS uses the same testing procedure as any other thermistor (Coolant temp. sensor, IAT, etc). This type of sensor (2-wire) can also be cleaned using carburetor cleaner.

- Disconnect the PFS connector.
- Turn the ignition key on.
- Using a voltmeter measure the voltage between battery positive and the PFS ground wire. Battery voltage should be seen (tests the PFS ground circuit).
- With the voltmeter measure the voltage across battery negative and the PFS positive wire from the ECM relay. Battery voltage should be seen again. (tests the PFS power feed circuit).
- Using the voltmeter measure the voltage across the PFS ground and battery feed wires. Battery voltage should be seen. (double-checks PFS ground and power feed circuit integrity).

If these diagnostics steps check OK, then the PFS is probably at fault. As an extra step perform the following.

- Disconnect the PFS vapor hoses. Take extreme care to crimp the EVAP vapor hoses when disconnecting the PFS.
- If necessary start the engine. (some ECMs need to see an RPM signal to provide a PFS scanner PID).
- Using shop air, introduce a small flow if air into the PFS sensor and monitor the PFS signal with a scanner or voltmeter. Voltage should climb up from around 1.5 volts to almost 5.00 volts. A cleaning procedure similar to the MAF sensing filament cleaning can be performed if the PFS is unresponsive, otherwise replace the PFS.

As a final note, be aware of the fact that the 3-wire PFS sensor is somewhat rare. If you get to diagnose one of them consider yourself lucky, **so the next day get yourself on a plane to Vegas because you'll definitely be a winner.**

WIDE-RANGE AIR FUEL RATIO SENSOR (AFR) DIAGNOSTICS STRATEGY

THEORY OF OPERATION

The early introduction of the oxygen sensor came about in the late 1970's. Since then Zirconium has been the material of choice for its construction. The Zirconium O2 sensor, as we all know, produces its own voltage, which makes it a type of generator. The generated varying voltage shows up on the scope as the familiar 1 Hz sine wave, when in **close loop**. **The actual voltage that is generated is the difference between the O2 content of the exhaust and that of the surrounding ambient air**. The stoichiometric air/fuel ratio or the mixture of air-to-fuel equal to 14.7:1 is the best mixture ratio for gasoline engines. At this ratio, the combustion process happens with the most power being generated and the least amount of emissions being produced. At a stoichiometric air/fuel ratio (14.7:1), the generated O2 sensor voltage is about 450 mV. The ECM recognizes a rich condition above the 450 mV level and a lean condition bellow it. Therefore, these sensors do not care about the air/fuel ratio above or bellow stoichiometry or 14.7-parts-of-air to 1-part-of-fuel. It is for this reason that the Zirconium O2 sensor is called a **"narrow band"** O2 sensor.

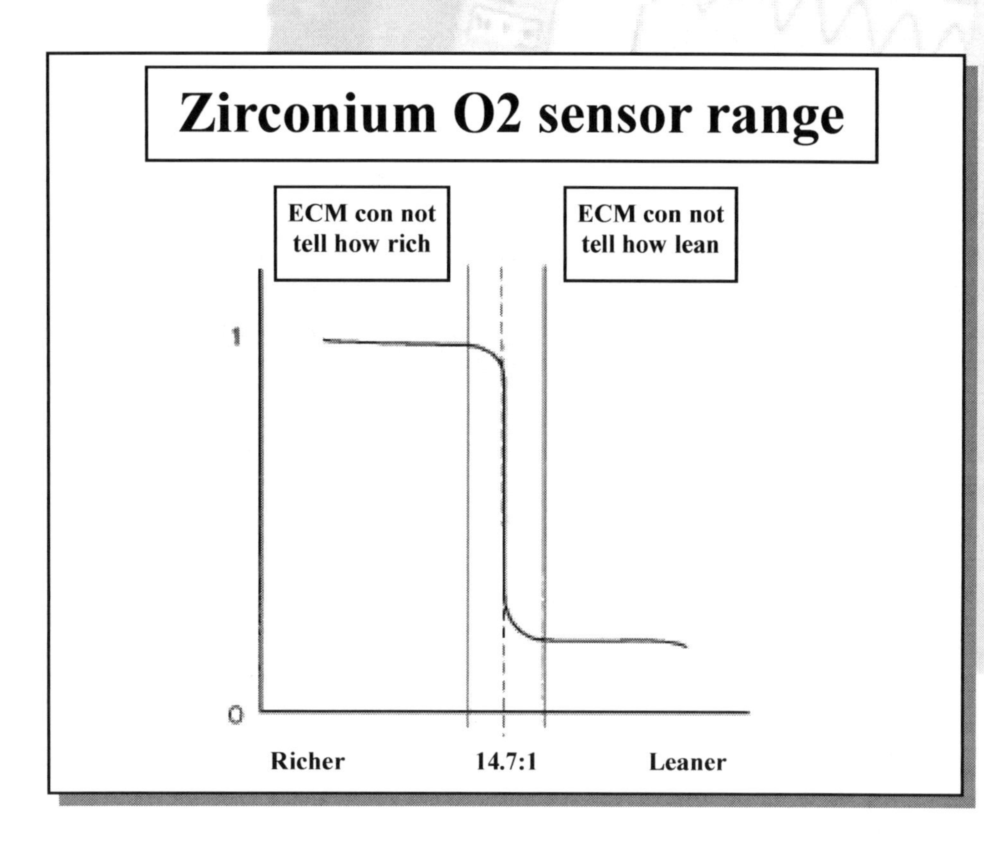

Fig 1 – Zirconium O2 sensor. This chart shows that bellow or above stoichiometry (14.7:1) the O2 sensor simply can not provide a reading at all. The ECM is totally blind to these wider A/F ratios.

The Titanium O2 sensor was used throughout the late 1980's and early 1990's on a limited basis. This sensor's semiconductor construction makes its operation different than the Zirconium O2 sensor. Instead of generating its own voltage, the Titanium O2 sensor's electrical resistance changes according to the exhaust oxygen content. When the air/fuel ratio is rich, the resistance of the sensor is around **950 Ohms** and more than **21 K-Ohms** when the mixture is lean. As with the Zirconium sensor, the Titanium O2 sensor is also considered a narrow-band O2 sensor.

As mentioned before, the main problem with any narrow band O2 sensors is that the ECM only knows that the mixture is slightly richer or leaner than 14.7:1. The ECM has absolutely no idea as to the operating A/F ratio outside the stoichiometric range. In effect it only knows that the mixture is richer or leaner then stoichiometry. An O2 sensor voltage that goes lower than 450 mV will cause a widening of injector pulse and vise-versa. The resulting changing or cycling fuel control (closed-loop) O2 signal is what the technician sees on the scope when probing at the O2 sensor signal wire.

The newer **"wide band"** O2 sensor solves the narrow sensing problem of the previous Zirconium sensors. These sensors are often called by different names such as, continuous lambda sensors, AFR (air fuel ratio sensors), LAF (lean air fuel sensor) and wide range O2 sensor. Regardless of the name, the principle is the same, which is to put the ECM in a better position to control the air/fuel mixture. In effect, the wide range O2 sensor can detect the exhaust's O2 content way bellow or above the perfect 14.7:1 air/fuel ratio. Such control is needed on new lean burning engines with extremely low emission output levels. The tighter emission regulations are actually driving this newer fuel control technology and in the process making the systems much more complex and difficult to diagnose.

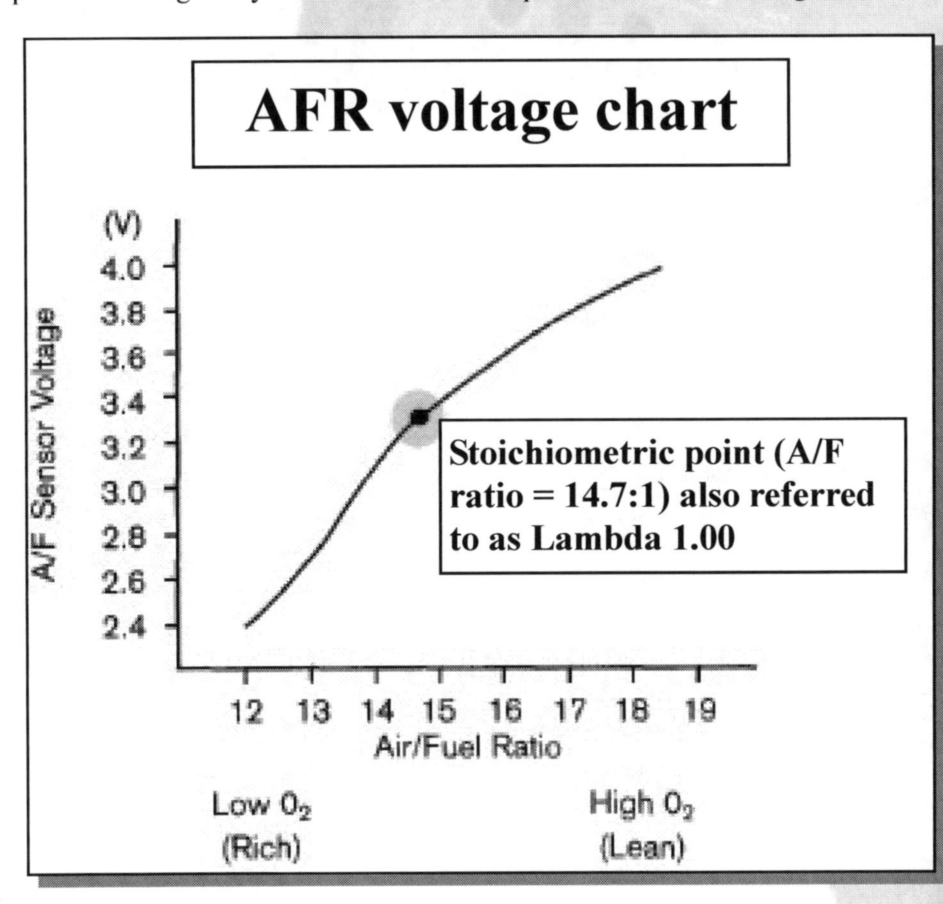

Fig 2 – Typical Toyota AFR sensor voltage chart. The voltage seen here is derived from the ECM detection circuit and NOT a directly measured value. Each manufacturer uses it own AFR sensor biasing voltage.

The wide range O2 sensor looks similar in appearance to the regular Zirconium O2 sensor. Its inner construction and operation are totally different, however . The Wide band O2 sensor is composed of a dual inner layer called **"Reference cell"** and **"Pump cell"**. The ECM's AFR sensor circuitry always tries to keep a perfect air/fuel ratio (14.7:1) inside a special **monitoring chamber (Diffusion Chamber or pump-cell circuit) by way of controlling its current**. The AFR sensor uses dedicated electronic circuitry to set a pumping current in the sensor's pump cell. In other words, if the air/fuel mixture is lean, the pump cell circuit voltage momentarily goes low and the ECM immediately regulates the current going through it in order to maintain a set voltage value or stoichiometric ratio inside the diffusion chamber. The pump cell then discharges the excess oxygen through the **diffusion gap** by means of the current flow created in the pump-cell circuit. The ECM senses the current flow and widens injector pulsation accordingly to add fuel.

If on the other hand the air/fuel mixture goes rich, the pump cell circuit voltage rapidly climbs high and the ECM immediately reverses the current flow polarity to readjust the pump cell circuit voltage to its set stable value. The pump-cell then pumps oxygen into the monitoring chamber by way of the reversed current flow in the ECM's AFR pump-cell circuit. The ECM detects the reversed current flow and an injector pulsation-reduction command is issued bringing the mixture back to lean. Since the current flow in the pump cell circuit is also proportional to the oxygen concentration or deficiency in the exhaust, it serves as an index of the air/fuel ratio. The ECM is constantly monitoring the pump cell current circuitry, which it always tries to keep at a set voltage. For this reason, the techniques used to test and diagnose the regular Zirconium O2 sensor **can not** be used to test the wide band AFR sensor. **These sensors are current devices and do not have a cycling voltage waveform**. The testing procedures, which we will go into further along, are quite different from the older O2 sensors.

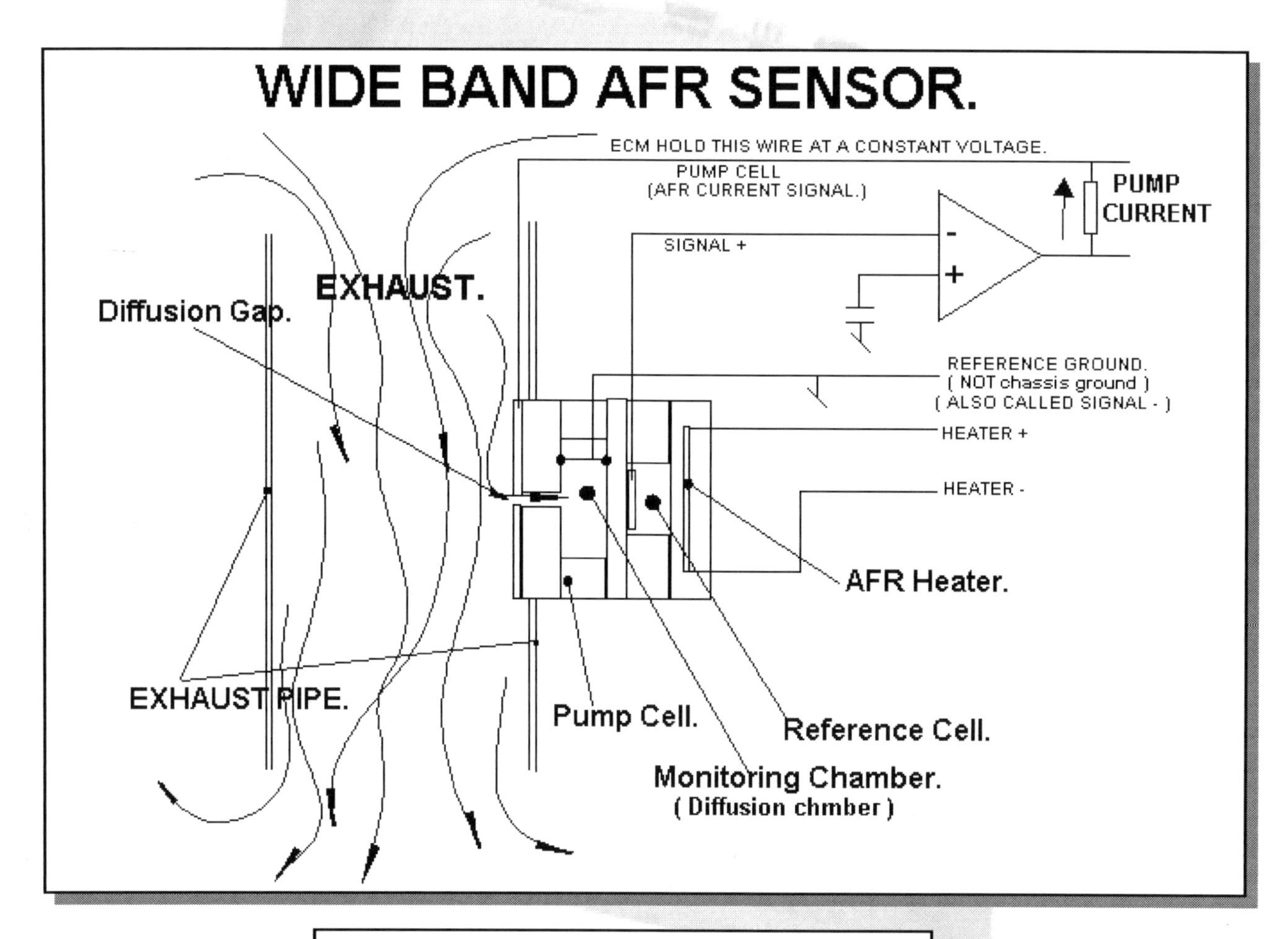

Fig 3 – Construction of a wide range AFR sensor.

NOTE

Whenever the air/fuel mixture is exactly at stoichiometry (14.7:1) there is no current flow through the AFR sensor. This is precisely what the ECM tries to do with the AFR signal. A properly operating engine will always have very close to 0.00 mA of current flow. The ECM commands more or less injector open time to try and keep the AFR sensor as close as possible to 0.00 mA. A rich mixture will produce a negative current flow and a lean mixture a positive current flow. The actual AFR current flow is extremely small and for this reason, the AFR sensor signal should be monitored with a scan tool.

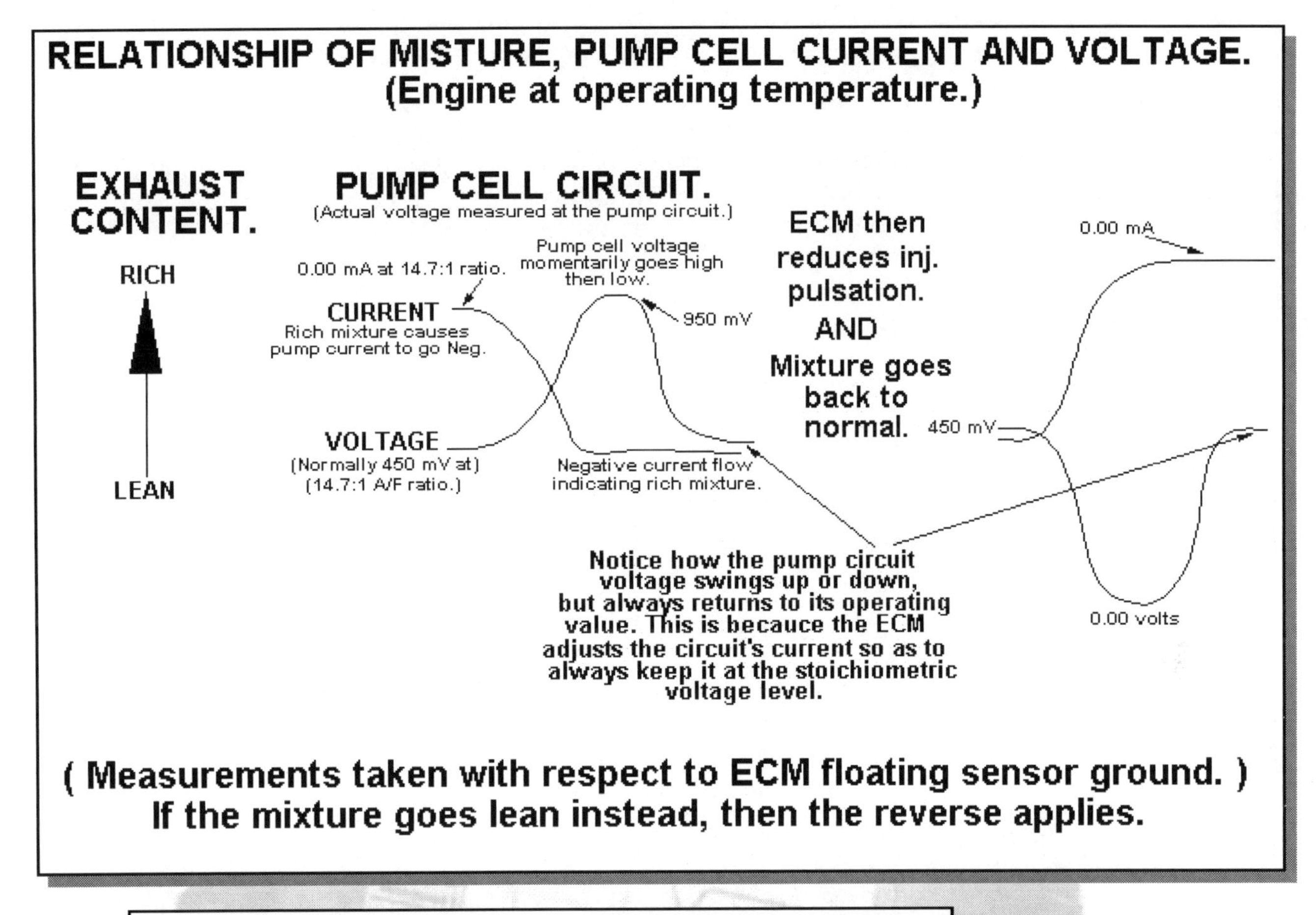

Fig 4 – Relationship between pump current, voltage and mixture.

The AFR sensor operation can be thought of as being similar to the hot wire MAF sensor. But, instead of a MAF hot wire, the ECM tries to keep a perfectly stoichiometric air/fuel ratio inside the monitoring chamber by varying the pump cell circuit current. The sensing part, at the tip of the sensor, is always held at a constant voltage (depending on manufacturer). If the mixture goes rich, the ECM will adjust the current flowing through the sensing tip or pump cell circuit until the constant operating voltage level is achieved again. The voltage change actually happens very fast. The current flow through the pump circuit also **pushes along the Oxygen atoms** either into or out of the diffusion chamber (monitoring chamber) which restores the monitoring chamber's air/fuel ratio to stoichiometry. Although the ECM varies the current, it tries to maintain the pump circuit at a constant voltage potential. As the ECM monitors the varying current, a special circuit (also inside the ECM) converts the current flow into a voltage value and passes it on to the serial data stream as a scanner PID. **This is why the best way to test an AFR sensor's signal is by monitoring the voltage conversion circuitry, which the ECM sends out as an AFR-voltage PID.** It is possible to actually monitor the actual AFR sensor varying current, but the changes are very small (in the low mA range) and difficult to monitor. A second drawback to a manual AFR current test is that the actual signal wire has to be cut or broken to connect the amp-meter in series with the pump circuit. Today's average clamp-on amp-meter is not accurate enough at such a small scale. For this reason, the easiest (but not the only) way to test an AFR sensor is with the scanner.

AFR voltage signal-to-Pump cell current chart. (Waveform taken at W.O.T. snap.)

At this point the A/F ratio is at 14.7:1, the pump cell current is at 0.00 mA and the scanner detection circuit voltage is at the mid-point level (in this case is a Toyota AFR sensor which is 3.3 volts).

As the throttle is snapped open, the ECM goes into fuel enrichment (rich mixture). This causes the ECM to apply a negative current flow to the AFR's pump cell circuit in order to keep the AFR diffusion chamber (tip of the sensor) at stoichiometry. The negative current flow is seen by the ECM current detection circuit and passed on to the data stream as a decreasing voltage data PID.

This is the point of maximum positive current flow from the ECM to the pump cell circuit (lean mixture). The ECM detection circuit registers the positive current flow as a scanner high-voltage PID.

This point shows the ECM increasing pulsation. It is trying to correct for the lean condition created by the injector cut-off.

At this point the ECM has stabilized the A/F mixture and the AFR pump cell circuit is back to 0.00 mA or an A/F ration of 14.7:1.

Maximum fuel enrichment attained by the throttle snap action. Pump cell circuit current at maximum (negative flow). At this point the ECM will start to correct the rich mixture and decrease injector pulsation (injector cut-off).

NOTE: The ECM varies the current in order to keep the pump cell at a constant voltage level at all possible conditions (stoichiometry).

Fig 5 – AFR signal-to-current analysis.

NOTE

Some diagnostics literatures suggest testing the AFR sensor by goosing the throttle and monitoring the actual voltage. On a good sensor the voltage will snap down-and-up and then go back to its normal level because the ECM will immediately adjust the current to maintain the constant operating voltage. DO NOT use a multi-meter in voltage setting to test the AFR sensor. The only voltage reading that should be used is the ECM's interpreted voltage value that is displayed as a scanner PID from the pump-current detection circuit.

Another major difference between the wide range AFR sensor and a Zirconium O2 sensor is that it operates at above 1200 Deg. F (600 C). On these units the temperature is very critical and for this reason a special pulse-width controlled heater circuit is employed to precisely control the heater temperature. The ECM controls the heater circuit.

The wide operating range coupled with the inherent fast acting operation of the AFR sensor puts the system always at stoichiometry, which reduces a great deal of emissions. With this type of fuel control, the air/fuel ratio is always hovering close to 14.7:1. If the mixture goes slightly rich the ECM adjusts the pump circuit's current flow to maintain the set operating voltage. The current flow is detected by the ECM's detection circuit, with the result of a command for a reduction in injector pulsation being issued. As soon as the A/F mixture changes back to stoichiometry, because of the reduction in injector pulsation, the ECM will adjust the current respectively. The end result is NO current flow (0.00 Amps) at 14.7:1 A/F ratio. In this case a light negative hump is seen on the Amp-meter with the reading returning to 0.00 almost immediately. The fuel correction happens very quickly.

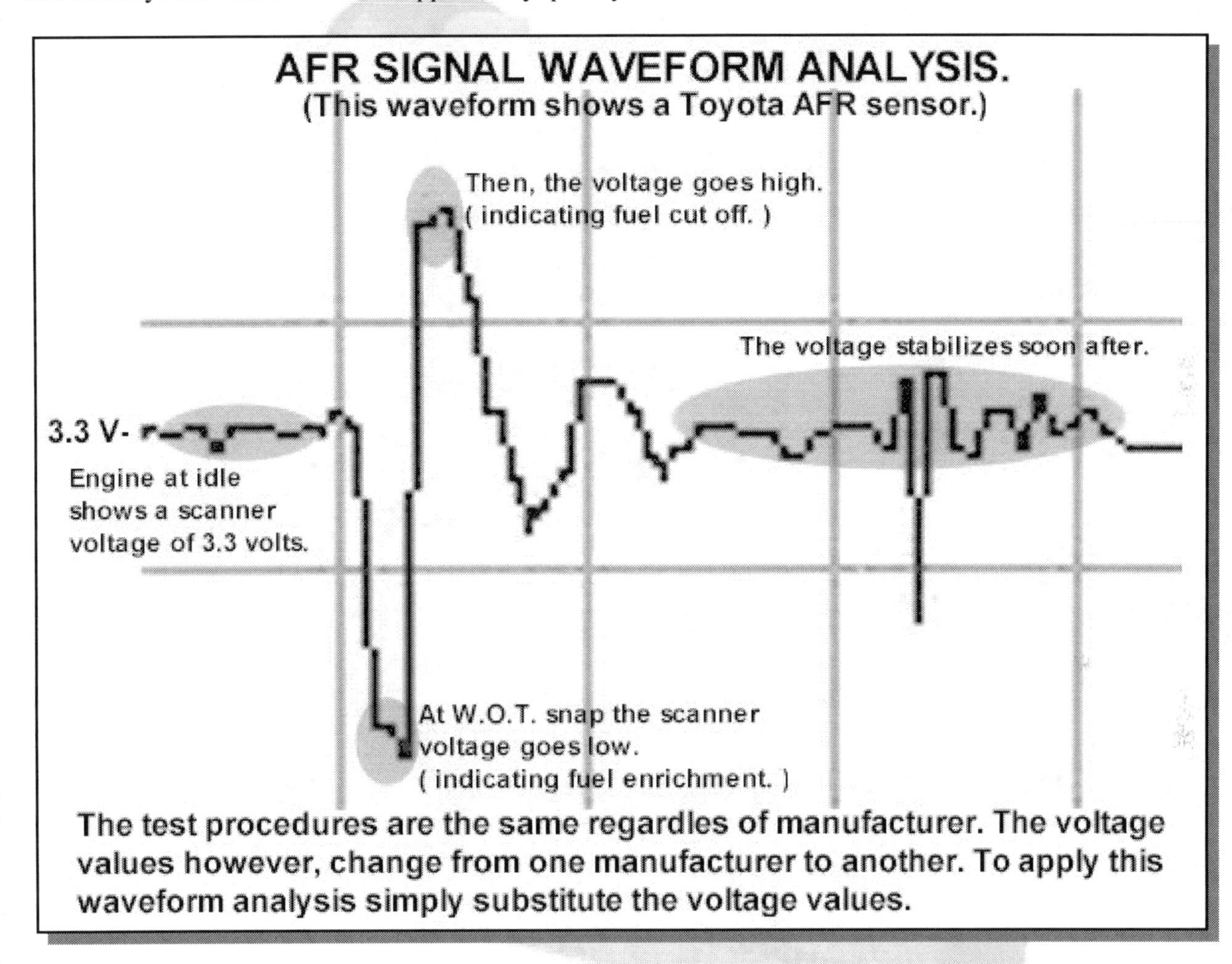

Fig 6 – AFR signal showing ECM response to a W.O.T. snap action.

Toyota among others has always been a strong supporter of wide-range AFR sensor technology. The OBD II regulation calls for an O2 sensor voltage range from 0.00 to 1.00 volt. In order to meet the OBD II regulation, Toyota rearranged the AFR sensor PIDs (from the detection circuitry) by dividing their original OEM PID value by 5. The newer generic OBD II AFR sensor PID ranges between 0.48 (rich) and 0.80 (lean).

NOTE

The AFR's pump-current detection circuit voltage range is the opposite of the regular Zirconium O2 sensor. With the AFR sensor, the lower the voltage value the richer the mixture, and the higher the voltage value the leaner. The OBD II generic AFR PID is called air/fuel ratio sensor and NOT O2 sensor.

The following table gives the values of the Toyota OEM PID, generic OBD II and the actual air/fuel ratio value.

AFR voltage OEM PID	Air/Fuel Ratio	OBD II PID (OEM value Divided by 5)
2.40 volts RICH	12.00 A/F	0.48 volts
2.50 volts	12.30 A/F	0.50 volts
2.60 volts	12.60 A/F	0.52 volts
2.70 volts	12.90 A/F	0.54 volts
2.80 volts	13.20 A/F	0.56 volts
2.90 volts	13.50 A/F	0.58 volts
3.00 volts	13.80 A/F	0.60 volts
3.10 volts	14.10 A/F	0.62 volts
3.20 volts	14.40 A/F	0.64 volts
3.30 volts	14.70 A/F (Stoichiometric ratio)	0.66 volts
3.40 volts	15.30 A/F	0.68 volts
3.50 volts	15.90 A/F	0.70 volts
3.60 volts	16.50 A/F	0.72 volts
3.70 volts	17.20 A/F	0.74 volts
3.80 volts	17.80 A/F	0.76 volts
3.90 volts	18.40 A/F	0.78 volts
4.00 volts LEAN	19.00 A/F	0.80 volts

Table 1 – Toyota AFR sensor OEM PID-to-OBD II generic PID (with A/F ratios).

The following summarizes the wide-range AFR sensor operation. The Toyota AFR sensor is used here as an example, since the operating voltages change from one manufacturer to another.

- The AFR sensor operates at a much wider air/fuel ratio detection range. Hence the name wide range.
- The AFR sensor provides the ECM with a signal value throughout a broad (wide) range of air/fuel ratios.
- The ECM current detection circuit voltage (scanner PID) is totally the opposite of a regular Zirconium O2 sensor. The higher the voltage, the leaner the mixture and vise-versa.
- The detection circuit voltage signal (scanner PID) output is proportional to the current flow applied by the ECM to the pump cell circuit (to keep the operating voltage) and an indicator of the air/fuel ratio.
- In AFR sensor fuel control systems, the ECM can more accurately measure the actual air/fuel ratio on a wider scale. This allows the ECM to adjust to stoichiometry much faster.
- With AFR sensor systems, the ECM does not cycle (rich/lean) as in the older Zirconium type O2 sensor. The output bias or pump cell circuit current detection voltage is fairly stable.
- With the mixture at 14.7:1, the AFR sensor pump cell circuit current flow is 0.00 mA.
- The **pump cell** circuit current flow **changes polarity (by polar).**
- A rich mixture produces a negative current flow in the pump cell circuit.
- A lean mixture produces a positive current flow in the pump cell circuit.
- Because the current can flow in either direction, **the AFR's ground is NOT chassis ground.** The AFR sensor uses a floating or ECM ground, which could be held at a specific voltage level above chassis ground (according to the manufacturer). Some manufacturers call this circuit **(Signal -).**
- The actual pump cell circuit current flow pushes Oxygen atoms into or out of the diffusion chamber, depending on the direction of the current flow.
- The **detection circuit** always monitors the direction of the current flow and how much of it is flowing.
- Toyota AFR systems show an AFR PID of 3.30 volts at 14.7:1 A/F ratio. Each manufacturer uses a different PID voltage value to signal the stoichiometric point. Toyota also divides its OEM PID by 5 in order to arrive at an OBD II compliant voltage value.
- The leaner the mixture, the higher the detection circuit voltage value (scanner PID). The richer the mixture the lower the detection circuit voltage value (scanner PID).
- The ECM tries to maintain a **stable voltage level** across the AFR's sensing tip or **pump cell circuit**.
- The AFR voltage reading on the scanner is not the actual voltage across the AFR sensor pump cell. The AFR detection circuit (inside the ECM) generates the scanner PID voltage data from the pump cell current flow. **The pump cell voltage is kept at a stable value by the ECM.**
- Wide-range AFR sensors are **current devices** and do not put out an **actual voltage** for their signal.
- The current output signal flowing through the AFR circuit is in the mA range and can not be measured with a clamp-on amp-meter.
- The same factors that affect the Zirconium O2 sensor also affect the AFR sensor (contamination, vacuum leaks, EGR failure, heater failure, etc).
- The AFR's heater operation is very critical to the sensor operation. These sensors operate at a much higher temperature than Zirconium sensors.
- The AFR heater is pulse-width modulated by the ECM to maintain a stable temperature.
- The AFR sensor heater is usually ON (pulsing) under normal driving conditions.
- **The AFR heater carries more current because of the higher temperatures necessary**. For this reason the connections are more critical so as to avoid resistance in the circuit.
- **The AFR heater circuit carries up to 8 Amps compared to the Zirconium O2 sensor at 1.5 to 2 Amps**

These sensors also have the added advantage of being able to have the fuel control system adjust to any desired air/fuel ratio **other than 14.7:1 (Stoichiometry) or lambda 1.** This option is especially important in new fuel control concepts such as **lean-burn engines**, where the engine's fuel control changes at cruising speeds from 14.7:1 to a much leaner 19.0:1 or even higher. The result is tremendous reduction in emissions and fuel consumption. It is also worth stating that these leaner engines require special catalytic converter units capable of reducing the considerable amounts of NOx generated at such leaner (high temperature) mixtures.

COMPONENT TESTING

The two more prevalent wide-range AFR sensor system manufacturers are Honda and Toyota. We will use Toyota in this section for explanation purposes. However, the testing procedure is always similar. The only changes will be in the biasing-voltage, which changes from one manufacturer to another. The basic operation is the same. Always learn the system before proceeding to further diagnostics. AFR sensors are also starting to appear in an increasing number of makes and models. It is expected that within the next decade most systems will be of this type.

The best way to test the AFR sensor operation is with a scan tool. With that in mind, the use of a graphing-software is highly recommended. This will ensure the quick recognition of the sensor's operating parameters much faster than simply looking at the numbers. The human brain can process **graphical information** much better than raw data numbers.

SENSOR TESTS:

- First, determine that there are no mechanical or air/fuel density problems (vacuum leaks, clogged air or fuel filter, ignition timing, stuck EGR, etc).
- Determine that the AFR sensor bias voltages are within specs. Using a voltmeter, disconnect the AFR sensor and back probe the sensor's signal wires. Probe between ground and one of the signal wires, then to ground and the other signal wires. The bias voltages should be measured to ground. Measure the bias voltages and compare to specs. The signal + wire is usually the pump cell signal circuit while the signal – is the ECM provided floating ground. **(Note: Some manufacturers are using an AFR sensor with 6 wires. Two heater wires, signal +, signal – and an extra two set of wires, which are the pump cell current signal and the calibration current wire.**
- Perform a W.O.T. snap test. The scan tool AFR voltage should reach a full low (rich mixture) voltage potential (Toyota – 0.48 volts (OBD II) or 2.40 volts (OEM), then a full high (lean mixture) voltage value (Toyota – 0.80 volts (OBD II) or 4.00 volts (OEM). ***(refer to fig. – 5).***
- If the AFR sensor did not pass the W.O.T. snap dynamic test above, suspect an open or a short circuit to the AFR current circuit. The AFR sensor has to react to the sudden W.O.T. snap test regardless of engine A/F conditions. A non-changing detection circuit is a very strong indication that the AFR circuit is down. Internal as well as external circuit faults are possible. To verify the fault (whether the fault is internal or external) disconnect the AFR sensor and either measure the continuity between the ECM AFR wires and the AFR connector or apply a varying voltage (O2 sensor simulator – 0.00 to 2.00 volts) to the AFR circuit. Look for a changing voltage in the scan tool's AFR data PID. **The actual amount of change is not important**, since this test simply checks for a changing response. This verifies that the circuit is not open or shorted, which would be indicated by NO change on the scanner AFR PID display.
- To perform a current response test of the pump cell circuit, simply break the pump cell circuit wire and connect a digital low-amp meter in between the broken circuit. Start the engine and warm up to operating temperature. Operate the engine at different conditions and compare to ***table 2***. (Table 2 is a general operating current value table, which may differ slightly from one manufacturer to another).

HEATER TESTS:

- Perform a voltage reading at the AFR heater power feed wire. Most AFR sensors a fed power through an ECM controlled power relay while the other side of the heater circuit is a pulse modulated to ground. (**Determines if the power feed voltage relay is operating properly).**
- Connect a low resistance lamp circuit (headlight) to the AFR heater circuit and start the engine. Verify that the lamp turns ON and OFF. **Note: The use of a low resistance lamp (headlight) is needed due to the fact that on some systems, the ECM constantly checks the heater resistance. On these systems, if a test light is used, the ECM simply shuts down the heater circuit, since a test light has a resistance of close to 20 Ohms and only draws about 300 mA of current, as opposed to the 8.00 Amps needed by the AFR heater.**
- Using a low amperage clamp-on meter, obtain a scope waveform from the AFR's heater circuit. The waveform should look similar to a pulsating ignition coil, with a series of current humps indicating a duty cycle (pulsing) controlled heater element. The highest part of the waveform should be within 6 to 8 Amps.

- A lack of a current hump on the scope's display points to the heater circuit not working. Disconnect the AFR sensor and measure the heater continuity. It should be close to 1.5 Ohm since the heater is almost like a straight through wire, which is why it is duty-cycle-controlled. If the heater circuit continuity reading shows an open circuit, replace the AFR sensor. Otherwise, perform a resistance check of the heater circuit wires between the sensor and the ECM/ Power feed circuit.
- A low current reading on the heater circuit indicates a high resistance fault in the heater circuit.
- A higher than 8.00 amps current draw from the heater circuit indicates a shorted heater inside the AFR sensor. Replace the AFR sensor and re-check.

Pump Cell Current	**Excess air in (Lambda)**	**Air/Fuel ratio**
- 0.005	**0.80 (RICH A/F MIXTURE)**	**11.7:1**
- 0.004	**0.84**	**12.3:1**
- 0.003	**0.89**	**13.0:1**
- 0.002	**0.91**	**13.3:1**
- 0.001	**0.95**	**13.9:1**
0.000 (NO CURRENT FLOW)	**1.00 (14.7:1—Stoichiometry)**	**14.7:1**
+ 0.001	**1.10**	**16.2:1**
+ 0.002	**1.25**	**18.3:1**
+ 0.003	**1.44**	**21.1:1**
+ 0.004	**1.74**	**25.2:1**
+ 0.005	**1.89**	**27.7:1**
+ 0.006	**2.00 (LEAN A/F MIXTURE)**	**29.4:1**

Table 2 – Pump cell current-to-A/F ratio reference data both in Lambda and A/F ratio.

NOTES:

Made in the USA
Charleston, SC
25 June 2015